Genetics
in the Wild

Genetics in the Wild

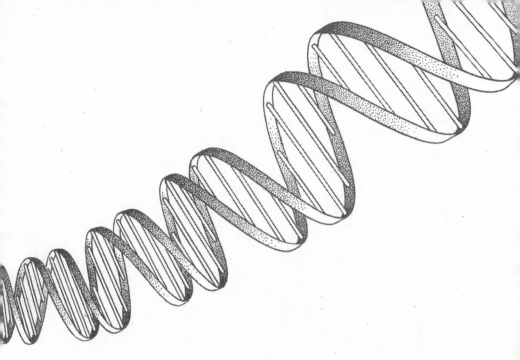

John C. Avise

Illustrations by Trudy Nicholson

SMITHSONIAN INSTITUTION PRESS
Washington and London

COPY EDITOR: Joanne S. Ainsworth
PRODUCTION EDITOR: Robert A. Poarch
DESIGNER: Janice Wheeler

Library of Congress Cataloging-in-Publication Data
Avise, John C.
Genetics in the wild / John C. Avise.
p. cm.
Includes bibliographical references (p.   ).
ISBN 1-58834-069-4 (alk. paper)
1. Animal genetics—Popular works. 2. Behavior genetics—Popular works.
3. Evolutionary genetics—Popular works. I. Title.
QH432 .A954 2002

591.3'5–dc21   2002017576

British Library Cataloguing-in-Publication Data is available

Manufactured in the United States of America
09 08 07 06 05 04 03 02   5 4 3 2 1

The paper used in this publication meets the minimum requirements of the American
National Standard for Information Sciences—Permanence of Paper for Printed Library
Materials ANSI Z39.48-1984.

For permission to reproduce the illustrations appearing in this book, please correspond
directly with Trudy Nicholson. The Smithsonian Institution Press does not retain
reproduction rights for these illustrations individually, or maintain a file of addresses for
photo sources.

Contents

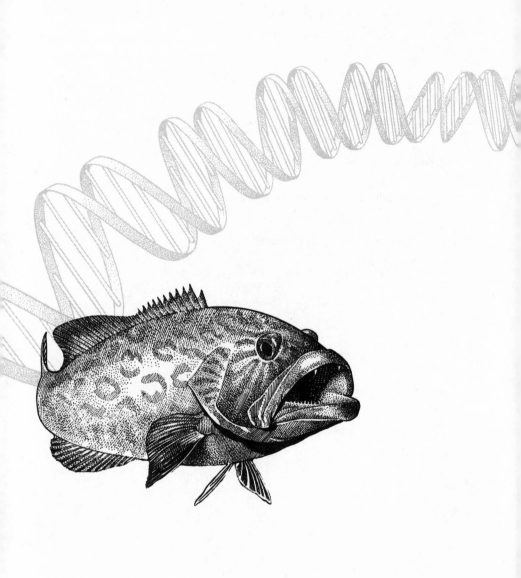

Preface

For as long as I can remember I have been awed by the diversity of nature. As a child, whether fishing in the lake behind my grandmother's house, collecting insects, or watching birds feed and fledge their young, I was driven by a desire to understand the fascinating creatures that share our planet.

When I was about five years old, my parents bought me a beautiful book that we loved to read together. It was entitled *Nature's Ways*, by Roy Chapman Andrews (New York: Crown Publishers, 1951). In a series of vignettes, it described the strange and wonderful adaptations of more than 100 unusual creatures: a fish with bifocal vision who sees above and below the water at the same time, a bird who finds food by walking underwater in torrential streams, a sea crab who glues a protective coat of sponge to his back for camouflage, a spittlebug who safely cloaks herself in a wet bubbly cocoon, and many more. Such wonders of nature surround us and are free for visual inspection by anyone who takes the time to notice.

People who study nature are called naturalists. Dr. Andrews worked for the American Museum of Natural History in New York, and I decided that when I grew up I wanted to be a professional natural historian, just like him. Many years later, in college, I chose a degree path in fisheries biology. In courses and field work I learned to identify all the fish species in Michigan and how to manage lakes and streams to help keep their populations healthy.

Then something unexpected happened. Some of my college instructors introduced me to another way of looking at the biological world: from a genetic perspective. Genes are fascinating objects of nature too, but they are far too small to be seen by the naked eye. Instead, their submicroscopic structures can be revealed only by sophisticated laboratory assays. Most geneticists focus on the detailed molecular structure of genes and how they function inside an organism. However, with my natural-history background, I began to ponder an altogether different kind of role for molecular data. Each animal or plant carries multitudinous small genetic variations that help to make it distinctive. Could this wealth of heritable information be tapped to learn more about the evolution and behaviors of organisms in the wild?

After completing graduate school, I became a university professor specializing in evolutionary genetics. This occupation has given me the best of two worlds—the freedom to be a natural historian and the opportunity to nurture an intellectual appreciation of life's evolutionary-genetic workings. During a thirty-year career, I have written hundreds of research articles describing hidden facets about the evolution, ecology, and behavior of creatures in nature as revealed by molecular genetic analyses. Many other scientists likewise have joined this emerging discipline, and some of us like to fancy ourselves as frontier naturalists, twenty-first-century style.

Research papers in molecular ecology and evolution now abound in professional journals. These articles, however, are technically demanding and can be difficult for general readers. In that respect, I miss those simple childhood days when my parents and I could share easily in nature's exciting stories. Yet, the new genetic sciences also have captivating tales to tell, and my intention here is to illustrate these in simple language. My broader goal is to inform and thereby instill in readers a sense of appreciation if not love for a marvelous natural world that, tragically, is coming under heavy siege from human activities.

This book's format and outlook, inspired by *Nature's Ways,* describes many fascinating and often spectacular discoveries about wild creatures from around the world, findings that could only have emerged from this new brand of genetic natural history. The essays are organized into topical sections but are written in a stand-alone style that should permit readers to delve at will into any that may pique their curiosity. Each genetic secret uncovered by molecular analysis was revealed through intense laboratory effort. Here, however, the spotlight will remain squarely on the organisms themselves in their natural environments and on what the genetic data have revealed about their intriguing lives.

Acknowledgments

The Pew Charitable Trust has supported my recent research and writing efforts in conservation biology. For the emotional and intellectual support of the Pew Fellows and the financial support of the foundation, I am deeply grateful. Helpful suggestions on all or parts of the manuscript were made by Michael Arnold, Betty Jean Craige, Cliff Cunningham, Andrew DeWoody, Paul Ehrlich, James Hamrick, Gene Helfman, Nancy Knowlton, Rodney Mauricio, Loren Rieseberg, Steven Schaeffer, Rytas Vilgalys, and Gerre Walker. More than 150 students in my evolutionary biology class at the University of Georgia provided a sounding board for an early draft of this work and also convinced me that in addition to its intended lay audience, the book could serve well as supplementary reading in undergraduate or graduate courses in which it might be desired to include a strong natural history component.

Special praise goes to Vincent Burke and DeEtte Walker for diligent editorial efforts. I am delighted that Trudy Nicholson, a gifted artist, was available to generate the book's superb drawings. Finally, I want to thank Mom and Dad for raising me with a deep and abiding appreciation for the natural world, and Joan and Jennifer for indulging my preoccupations with natural history and science.

Introduction

To a wildlife lover, nature can be like a living curio shop, chock-full of fascinating biological knickknacks. In this book we'll enter this enchanting emporium, rummage around for a genetic peek at some beguiling creatures, and thereby gain a sense of nature's bounty and the eclectic lifestyles of some of its more outlandish occupants. Each vignette raises and addresses some previously unanswered question about organisms in the wild, the new insights coming in each case from the focused study of genetic molecules.

DNA (deoxyribonucleic acid) is the primary genetic material of life, the coded molecular blueprint ultimately specifying the functional operations of all plants and animals. In any organism, the total amount of DNA-encrypted information is astronomical. Each human cell, for example, contains more than three billion nucleotides (the chemical building blocks of DNA) strung together in an epic biological script. Various small genetic passages (DNA sequences, or genes) in this voluminous text specify the tens of thousands of different kinds of protein molecules that actually conduct most of a cell's biochemical and physiological activities. Tens of thousands of other DNA sequences play as yet no known functional role in the cell, yet they too are historical outcomes of the evolutionary process.

The field of genetics focuses primarily on elucidating the functions of genes and their protein products, but there is another way to appreciate these wonderful molecules—as biological "markers." To the new breed of genetic naturalists, DNA and protein molecules are viewed as miniature hereditary tags or special genetic badges that Mother Nature herself has applied to every living thing. Properly read, molecular genetic markers provide a wealth of otherwise hidden information on the contemporary and historical lives of the animals and plants that house and transmit them.

For example, molecular genetic variation is so high in most species that scientists often employ these markers to distinguish each individual from all others. In other words, using appropriate laboratory assays, they can scan genes to identify and distinguish particular specimens (from even the smallest body part) much as store clerks use a scanner on computerized bar codes to identify shopping items. Most people are familiar with the use of "DNA fingerprinting" in courtroom cases, but they may be less familiar with the many applications in wildlife investigations.

Because genes are passed from generation to generation in specifiable ways, genetic assays also can be used to decipher an individual's close family ties. For example, a special kind of genetic material known as mitochondrial DNA (mtDNA) is transmitted almost exclusively via females. Thus, mtDNA's inheritance is nature's matrilineal analogue of patrilineal family-name "inheritance" in many human societies: just as human sons alone transmit their father's surname to the next generation, so do daughter animals pass their mother's mtDNA to their offspring. By contrast, most other genes (housed in the cell's nucleus rather than in its mitochondria) are transmitted via both genders. Genetic naturalists routinely use both such classes of molecular marker to identify a creature's parents and its other relatives, information that otherwise may be unavailable from field observations alone. Such knowledge on close kinship can shed much new light on a species' mating behaviors, reproductive activities, and dispersal abilities. Sometimes, even the bygone lifestyles of now-extinct organisms can be deduced through such evolutionary-genetic detective work.

Delving deeper into the past, every creature on Earth had parents who in turn had parents and so on back through time, eventually across millions of generations. Genetic assays of living organisms can help to plumb these evolutionary depths as well. The results, pictorially summarized as twigs and branches in the "tree of life," encapsulate the historical relationships (phylogenies) of living species and their ancestors. In the extreme, molecular genetic data have been used to explore some of the tree-of-life's most ancient limbs. In short, by properly reading life's evolutionary diaries recorded in DNA, biologists can unveil a wealth of natural-history secrets about living as well as extinct organisms, insights that could not have been gained merely by observing current-day plants and animals in the field.

In the last four decades, genetic markers provided by the various laboratory tools of molecular genetics (see Appendix) have illuminated the ecologies and lifestyles of thousands of species, ranging from bacteria to whales. Every organism has a fascinating tale to tell—there are no dull creatures once you imagine walking, swimming, or flying inside their skins, or envision how and when they evolved from shared ancestors. Thus, hundreds of compelling genetic stories could be told, and it was not easy to narrow the current list to just ninety-two essays. In choosing particular examples, I endeavored to include a diverse array of animals, plants, and microbes, as well as a wide range of behavioral, ecological, and evolutionary phenomena. These illustrate, but far from exhaust, the discipline.

Some of the discoveries were revolutionary, of broad significance to science; others merely offer tantalizing glimpses into the current or past lives of a few special species. Some of the genetic disclosures resulted from meticulous, planned detective work; others were serendipitous. All should appeal to readers curious about the workings of the natural world.

Genetics in the Wild

1

Some Evolutionary Oddities

Molecular markers have yielded some surprising discoveries about the historical relationships (phylogenies) of many intriguing microbes, plants, and animals. Molecular tools can be applied alike to organisms small and great, from tiny viruses to giant pandas. Genetic data can address biological mysteries across a full spectrum of evolutionary timescales—from genealogical connections among modern populations that separated just a few centuries ago, to extremely ancient branches in a tree-of-life.

Phylogenetic trees based on genetic markers yield historical backdrops for interpreting a host of topics ranging from behavioral evolution (How did some plants become meat-eaters?), to morphological questions (Why do king crabs have a lopsided rear end?), to physiological issues (How many times did algae establish close symbiotic relationships with corals?), to queries on evolutionary mechanisms (Did viruses arise from "jumping genes"?). The natural-history stories in this first chapter also will introduce a variety of special creatures from around the world, from snakes in southeast Asia, to giant tubeworms thousands of leagues beneath the Pacific Sea, to "living fossil" horseshoe crabs crawling along the muddy shores of Delaware Bay.

The Panda's Pedigree

Thanks in large part to Ling-Ling and her male partner Hsing-Hsing, giant pandas *(Ailuropoda melanoleuca)* are no longer complete strangers to many Americans. Before their recent deaths, these two endearing animals, a diplomatic gift from China, were housed in the National Zoo in Washington, D.C., where they were exemplary good-will ambassadors of their vanishing kind. Sadly, the giant panda is a highly endangered species. Fewer than 1,000 individuals are left in the wild, clinging to survival in their native alpine forests on the edge of the Tibetan Plateau in western China.

Dressed in fuzzy coats, with large black eyespots and Mickey Mouse ears, giant pandas look like enormous stuffed teddy bears. Indeed, many biologists suspected that these shy creatures were allied closely to real-life bears (family Ursidae). One early naturalist who promoted this notion was Père Armand David, a French missionary in China who in 1869 provided the first scientific description of these animals and named the species *Ursus melanoleucus* ("black-and-white bear").

Still, the giant panda displays many traits that are not at all bear-like. Unlike bears with their flashing canines, giant pandas have flattened teeth only, which they employ to munch their favorite food—fresh bamboo. Unlike regular bears, giant pandas don't hibernate, and they also have a unique and distinctive opposable "thumb" (really a modified wrist bone). Finally, instead of issuing a deep growl like a bear, the panda gives out a plaintive bleat, like a sheep. So, is the giant panda a bear or not?

Based on other lines of evidence, some scientists raised an alternative hypothesis—that the giant panda might not be allied closely to bears, but rather to raccoons (*Procyonidae*), well-known mammals that are thought to be even more distant evolutionary cousins of the true Ursidae. This is a classic kind of phylogenetic dilemma simply begging for examination using molecular genetic markers. To whom is the giant panda most closely related?

Many different kinds of laboratory assays have been employed on this question. The generally agreed-upon genetic picture to emerge is that the giant panda's ancestors were an early offshoot of the bear lineage, rather than in the direct line of descent leading to raccoons and their allies. Apparently, the proto-panda lineage split off from the proto-bear lineage about 20 million years ago, a scant 10 million years after the evolutionary lines eventually leading to bears and raccoons had themselves separated from a common ancestor. Thus, genealogically speaking, giant pandas proved to be just barely on the bear side (rather than the raccoon side) of the mammalian family tree.

Meat-Eating Plants

Plants are supposed to be subservient to us animals, providing food, shelter, and atmospheric services and adding beauty to our lives. But don't tell that to an impressive cadre of herbaceous species who would rather eat the zoological world than indulge it. I'm referring to the carnivorous plants.

Meat-eating plants come in hundreds of species and a diversity of forms. There are species of trap design that snap shut when entered by an unsuspecting fly. Others have the shape of an upright vase or pitcher, whose inner walls, lined by down-curved hairs, prevent the escape of any insect that may have tumbled inside. There are flypaper plants that act as their name implies, secreting sticky substances to hold and then help digest their prey. Still other meat-eating plants use small, aquatic suction-like devices (known as bladders) to capture insects.

Charles Darwin himself was impressed by the varied dining adaptations of carnivorous plants, and in 1875 published an entire book on the topic. Understandably, he adopted the conventional taxonomic wisdom of the time—that such diverse meat-eating groups as snap-trap plants, pitcher plants, sticky sundews, flypaper species, and the aquatic bladderworts had each evolved independently from unrelated, noncarnivorous ancestors.

Any plant's evolutionary transition to carnivory, however, must involve a great host of morphological, biochemical, and physiological alterations. Not only must the plant evolve novel devices to attract, retain, and kill its meaty quarry, but it also must devise means to digest, absorb, and metabolize the animal flesh it consumes. The underlying complexity of this suite of abilities suggests that the evolutionary transition to carnivory by plants might be extremely difficult, and perhaps rare. Hence, a competing hypothesis is that all carnivorous plants stemmed from only one or a few common ancestors that first made the challenging evolutionary transition to a meat-eating habit.

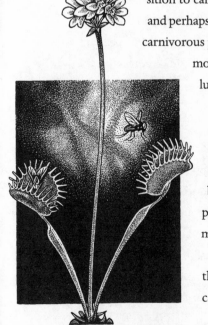

On how many independent evolutionary occasions did carnivory arise in the history of flowering plants? To answer this question, researchers generated a phylogenetic tree based on DNA sequences for more than 100 plant species, including representatives of all the major meat-eating types.

These molecular genetic analyses showed that Darwin and the earlier taxonomists basically were correct—different kinds of carnivorous plants tend to be scattered about on several different branches of the phylogenetic tree of plant life. For example, bladderworts had a separate evolutionary origin from pitcher-shaped plants, which themselves arose independently from noncarnivorous forms on at least three separate occasions. The molecular phylogeny also indicates, however, that some of the finer adaptations for carnivory, such as details of the glandular apparatus in many flypaper plant types, originated only once during the evolutionary process. Personally, I'm just glad that all the carnivorous plants remained small.

Venomous Vipers and Their Toxins

Many snake species secrete potent venoms that contain neurotoxins for immobilizing prey, or digestive enzymes that help the snake both to kill its victim and to process its newly acquired meal. Venom also can serve the snake indirectly, by deterring predators. Aversion to snakes may be learned, or it may be quite instinctual (as some people with ophidiophobia, a heightened aversion to serpents, may wish to attest).

In southeast Asia, the Malayan pit viper *(Calloselasma rhodostoma)* is the leading cause of venomous snakebites for humans. The powerful digestive enzymes in its venom can kill or permanently deform unlucky victims by causing massive tissue death. Interestingly, however, the potency of the snake's venom varies greatly from one geographic locale to another, as judged by the varied symptoms displayed by bite victims. Much of this variation in response appears to be related to the molecular genetic composition of the venom itself.

In nature, adult Malayan pit vipers eat amphibians, other reptiles, birds, and small mammals. However, their menus differ considerably from place to place, as scientists discovered by examining the feces and stomach contents of hundreds of snake specimens. For example, one population in western Thailand eats cold-blooded reptiles exclusively, whereas a nearby population in south-central Thailand dines primarily on warm-blooded animals.

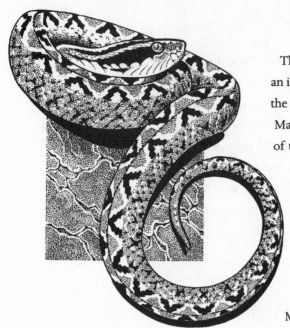

These observations raised an interesting question: Are the venomous properties of Malayan pit vipers a legacy of the snakes' evolutionary past, or are they related to recent selection pressures stemming from the serpents' varied diets? Analyses of DNA sequence data show that the Malayan pit viper species consists of three distinctive historical genetic lines, each confined to a particular geographic region: western Thailand; eastern Thailand plus Vietnam; and Malaysia, Java, and Thailand's Rayong Province. These evolutionary lineages do not, however, align with geographic differences in venom composition and potency. Instead, the properties of the venom are highly correlated with the snakes' local diets.

These findings suggest that venom composition in Malayan pit vipers evolves extremely rapidly, because the varied toxin types differ considerably within each of the major historical lineages within the species. Thus, quite likely, different populations of the Malayan pit viper quickly evolve venoms appropriate for subduing and digesting locally available food items. The different taxonomic groups (or taxa) that make up the prey are known to vary markedly in their susceptibility to snake toxins, so it is entirely plausible that location-specific features of the venom are finely tuned by natural selection to meet the snakes' needs, as are behavioral and morphological features of these slithery predators. These studies illustrate how molecular markers of the present can unearth both the tempo and mode of genetic processes of the recent evolutionary past.

Giant Tubeworms

What large animal has no mouth, no guts, and no digestive system? It's the famous giant tubeworm, *Riftia pachyptila*, a glorious creature in scientific circles. These impressive beasts, resembling turgid sausages reaching nine feet in length, are charismatic poster children of the incredible "hydrothermal vent communities" recently discovered in the oceanic depths.

The tubeworm doesn't need to eat in the conventional way because it has established a symbiotic relationship with "chemosynthetic" bacteria. These deep-sea microbes produce complex organic carbon compounds by chemical oxidation, a strikingly different procedure from the light-driven reactions of plant photosynthesis that support food chains elsewhere on Earth. Each tubeworm literally is stuffed with zillions of these bacteria, whose waste products nourish their tubular host. In return, the worm's bloodstream carries needed nutrients to the bacteria, so everyone stays well nourished in this pitch-black domain thousands of leagues beneath the sea.

Hydrothermal vent communities were discovered first in the late 1970s by the submersible research vehicle *Alvin*. Additional other-worldly animals inhabiting these deep-sea ecosystems include the aptly named Pompeii worm, spaghetti worm, and dandelion worm, as well as newly described species ranging from clams to copepods, limpets, crabs, barnacles, and shrimp. All these specialized animals live perilously close to superheated waters issuing from volcanic rifts in the seabed, and they owe their existence directly or indirectly to the chemosynthetic services of microbes.

On the seafloor, fields of hydrothermal vents, confined to the thin margins between Earth's grinding crustal plates, are few and far between, like small isolated islands. Hydrothermal vents are also ephemeral, some lasting only a few decades. Accordingly, it is suspected that most hydrothermal vent species with broad geographic distributions (including giant tubeworms) must have in their life cycle a highly dispersive stage—free-swimming larvae.

Many questions surround the "cool" creatures found in hydrothermal vent communities. Two that have been addressed using molecular ge-

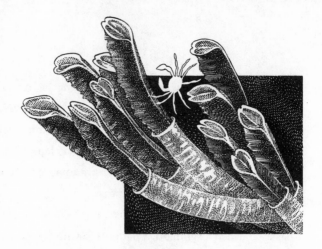

netic markers are as follows: To whom are the giant tubeworms most closely related evolutionarily? and, Are geographic populations in close genetic contact by virtue of occasional larval dispersal between different vent communities?

Based on comparisons of DNA sequences and other data, tubeworms are not typical "worms" at all (such as familiar earthworms, flatworms, or roundworms). Instead, they are a distinctive evolutionary group, at best only distantly related to the wiggly creatures dug from our gardens to bait fish hooks. Accordingly, tubeworms sometimes are placed into their own unique phylum, Pogonophora. This obscure group, of which scientists were completely unaware prior to 1900, is now believed to consist of several dozen species largely confined to deep oceans.

Another genetic issue concerns the population structure of *Riftia pachyptila*. From assays of proteins and DNA, it has been learned that giant tubeworm populations even on distant hydrothermal vents can be quite similar genetically. This implies that larval movements among them are recent or ongoing. Hydrothermal vent communities are arranged in discontinuous series of broken lines that stretch for thousands of miles along the boundaries between crustal plates. It seems doubtful that larvae would encounter a suitable site unless it was fairly close to where they were born. Thus, successive generations of tubeworms probably tend to move from one vent community to the next in stepping-stone fashion. Exactly how the microscopic tubeworm larvae manage to find suitable settling places in the vastness of the ocean, however, is just one of the remaining biological mysteries of this exciting deep-sea frontier.

Horseshoe Crabs

An ancient ritual takes place in Delaware Bay, the third week in May, at high tide. A female horseshoe crab plows slowly across the shallow sea-grass flats like a miniature armored tank. She is accosted by a squadron of male suitors, one of whom climbs onto her rounded shell and is dragged ashore, the others following close behind. With her amorous cadre in tow, the female crab ascends the beach, digs a shallow depression, and deposits several hundred eggs into the wet sand. The males do their duty by releasing sperm, and the fertilized eggs hatch a few weeks later into tiny babies, each less than half an inch long, that scurry back into the sea. There they will undergo more than a dozen molts, gradually growing into two-foot-long adults.

This is but one small scene in a long-running biological play whose cast can involve 10,000 mating horseshoe crabs packed along a mile of Delaware beach. Also in the theatrical troupe are gluttonous gulls and thousands of smaller shorebirds who have timed their global travels perfectly to attend this annual banquet of nutritious crab eggs. For many sandpipers, it is a critical fuel stop in their migrational journey from South America to breeding grounds in the high Arctic.

Horseshoe crabs that patrol marine coastlines today look essentially the same as those that wandered the seas more than 200 million years ago. There are four living species: *Limulus polyphemus* in the western North Atlantic, and three more look-alike species in the seas of southeastern Asia. Like its ancestors, each modern horseshoe crab has a

rounded hinged shell adorned with creases and bumps and housing two eyes that gather light and sense movement. Several spikes project backward from the rear of the shell, as does a fearsome-looking but harmless pointed tail. On the animal's unprotected underside are complex mouth parts for scavenging bits of food from the seabed, five pairs of jointed walking legs, and a series of gills stacked on top of one another like tattered magazines.

Because horseshoe crabs alive today closely resemble those registered in the fossil record for hundreds of millions of years, they are sometimes referred to as "living fossils" or as "phylogenetic relics." Does morphological constancy across vast spans of evolutionary time imply that these ancient creatures lack appreciable genetic variation? Have they failed to evolve genetically?

Molecular genetic assays of numerous proteins and DNA sequences have revealed that modern-day horseshoe crabs are unexceptional in levels and patterns of genetic variability, both within and across species. In other words, most of their genes give no indication of "aberrant" evolutionary-genetic rates or processes compared with those of other kinds of animals. Why then do these ancient arthropods show such morphological stasis? It's not because their genes can't or don't evolve. Apparently, horseshoe crabs long ago came up with a body plan that works well, and have simply stuck tenaciously to it. Yet, internally, their molecular genetic clocks have kept on ticking.

Returning to the scene in Delaware Bay, new actors have walked onto the stage. These are humans, there to harvest adult horseshoe crabs to be chopped for bait and drained of their bluish-green blood for pharmaceutical products. To support these commercial industries, horseshoe crabs are killed by the tens of thousands annually. Might this assault be a fatal blow to these creatures, finally causing their extinction? Unfortunately, this is an all-too-real possibility, so wiser conservation practices are needed.

However, don't underestimate these resilient beasts. Their kind has witnessed mountain ranges rise from the sea, and continents drift across the full face of the planet. Horseshoe crabs roamed the seas long before birds and mammals were more than a distant dream of Mother Nature, and they watched through their simple eyes as the dinosaurs came and went within a mere 100 million years. So, why should horseshoe crabs be particularly impressed by the latest challenge from an upstart creature, *Homo sapiens*? From the crab's perspective, our Johnny-come-lately species has inhabited the planet for only a brief instant—about one one-thousandth the evolutionary duration of these ultimate survivors.

A Tale of the King and the Hermit

Genetic markers can be employed to solve even the most unlikely of evolutionary mysteries. Here's an example involving the origins of a peculiar trait—a lopsided rear end—shared by some royal and peasant crustaceans.

To seafood lovers, king crabs from the deep cold waters of Alaska are a succulent delicacy. Morsels of tender meat await the diligent connoisseur who cracks apart the hard casing (exoskeleton) that completely covers these animals' foot-long legs and circular bodies. For their exceptional physical presence as well as delicious taste, it's little wonder that these are known as the king of crabs.

Hermit crabs, by comparison, are small unappetizing creatures that spend their unobtrusive lives grubbing around in shoreline mud. Unlike king crabs, hermits have a naked abdomen that makes them potentially highly vulnerable to predators. However, hermit crabs have an ingenious solution to this predicament. Each animal stuffs its coiled, naked little butt into the abandoned shell of a snail, leaving only its head and legs protruding from the aperture of the snail's former home. Hermits spend a great deal of time house hunting, searching for a vacant shell with just the right size, shape, and feel. As judged by the hermit crab's abundance and documented presence in the fossil record for more than 150 million years, this shell-inhabiting lifestyle has been a highly successful evolutionary strategy.

The coiled rear end of a hermit crab makes perfect sense, fitting nicely as it does into the coiled turret of an abandoned snail's shell. Oddly, however, free-living king crabs also have a similarly shaped derriere. Why in the world would king crabs also possess an asymmetrical, hermit-like hind end?

To answer this question, geneticists examined DNA sequences from hermit and king crabs and used the data to reconstruct the evolutionary tree for these species. Remarkably, the king crab lineage proved to be fully embedded within the broader family tree of hermit crabs. In other words, king crabs evolved from hermit crab ancestors, a process that apparently began a mere 13 million to 25 million years ago. Additional lines of evidence further support the notion that king crabs and hermit crabs are close evolutionary cousins. For example, their larvae are remarkably similar. Also, some of the five pairs of legs in both king and hermit crabs are greatly reduced in size, unlike the situation in many other taxonomic groups of crabs.

Thus, the asymmetrical rear end of king crabs is a historical legacy, an evolutionary leftover from their snail-shell-inhabiting ancestors. What originally prompted creatures in the king crab lineage to abandon the safety of snail shells? One reasonable hypothesis is that snail shells are relatively scarce in the deep-sea environments where king crabs live. Without the assurance of safe havens, selection pressures must have been strong indeed for protective exoskeletons of the crabs' own creation. In any event, it is now abundantly clear that the end product of this evolutionary pathway to king crab royalty had humble origins in a hermit's hovel.

Plantlike Animals Full of Algae

An adult staghorn coral *(Acropora cervicornis)* is a slender, branched animal that looks and grows like a plant. It began life as a tiny larva, bred from a fusion of egg and sperm, floating in the ocean like a seed. The larva then settled onto the reef to begin its sedentary existence. Cementing itself to a bit of rock, the little animal gradually proliferated into thousands of genetically identical polyps, each of which secreted about itself a hard wall of calcium. Each polyp occupies one small room in a residential structure that can grow several inches per year from each of several tips. Thus, the adult colony looks like a two-foot-high shrub, or the multi-pronged antlers of a prime buck deer (hence the coral's common name).

Staghorns and other corals resemble plants in another way as well—they undergo photosynthesis (the biochemical process by which nearly all plants produce food from carbon dioxide and water). Actually, the corals themselves don't photosynthesize, but special one-celled algae growing within their tissues do. These are known as dinoflagellates, or zooxanthellae. Unlike their free-living algal cousins, these symbiotic zooxanthellae are dependent on corals or other marine animals for suitable housing.

To a taxonomist, each such algal cell is a minuscule yellow-brown dot, and one pretty much looks like any other. Thus, many systematists placed symbiotic zooxanthellae into a single species, *Symbiodinium microadriaticum*. If all such zooxanthellae truly belonged to one species, this would be a catholic creature indeed because in addition to corals, these algae also inhabit sea anemones, giant clams, and a host of other invertebrates in the marine realm. Alternatively, perhaps there are many cryptic species

13

of animal-associated zooxanthellae, some perhaps confined to particular host species. This issue was attacked by molecular genetic analysis.

Despite their bland morphology, it turns out that zooxanthellae are a remarkably diverse group consisting of many species that often differ rather profoundly in genetic makeup. Indeed, some *Symbiodinium* species are as genetically divergent from one another as are different taxonomic orders of their distant cousins, the free-living dinoflagellates (which by comparison are a morphologically varied lot).

Furthermore, the genetic markers revealed that highly dissimilar zooxanthellae can be found in the same animal host, and conversely, that genetically similar zooxanthellae can inhabit taxonomically diverse animals. This lack of a close match between the phylogenies of the host animals and their algal houseguests implies that "host-switching" has been a common evolutionary phenomenon. Evidently, in the history of these plant-animal associations, the algae frequently have jumped from one host species to another.

The seas' coral reefs are currently being lost and degraded at an alarming pace. One obvious sign worldwide is "coral-bleaching," a stress-related phenomenon wherein corals lose their intimate algal associates, and then blanch and die. For months after its untimely death, each coral colony leaves behind a ghostly skeleton, a lifeless superstructure of calcium deposits that once supported a bustling condominium of interdependent plants and animals. These stark tombstones sprawled across the reef provide a disheartening testimonial to the declining health of the world's tropical oceans due to such stresses as global warming and coastal pollution.

The Bacterial Bounty Within

To several hundred species of bacteria and other microbes, the human body is home-sweet-home. These microscopic creatures inhabit our every nook and cranny, from outer skin to inner gut. Collectively, their numbers are enormous, with each of us harboring far more microbial specimens than there are humans on Earth. Some parts of our body are sparsely settled, such as the broad expanse of skin across our back (which must seem like a vast inhospitable desert from a bacterium's viewpoint). Other spots, such as our warm humid armpits, mouth, and intestines, are more like lush tropical rainforests, densely populated with bustling microbial communities.

Our microbial associates are lifelong companions, colonizing us shortly after birth via suckling and other normal human contacts, and remaining faithful until the day we die. Some are passive hitchhikers, merely along for the ride. Others pay their own way through the services they provide, such as assisting in our digestive processes. Occasionally, outsiders with more nefarious motives hop on board via a handshake, kiss, or sneeze. Depending on the microbe and our inherent susceptibility, these invaders can cause illnesses ranging from the common cold to some of the most deadly of human diseases.

As intimately tied to our lives and fates as these various microbes are, there is another category of "bacteria" that is embedded even more firmly within our very being, so integral a part of our essence that it is no longer possible to distinguish "them" from "us." These are the organelles (miniature "organs" within each cell) known as mitochondria. Not only do mitochondria inhabit our cells, but they also provide critical services upon which our lives depend. Namely, they act as miniature intracellular batteries, supplying much of the energy that our cells need to survive and reproduce.

Each mitochondrion in humans (and in nearly all other multicellular animals) contains a snippet of DNA about 16,000 letters (nucleotides) long. Close inspection of relevant portions of these genetic markers has revealed an incredible surprise. Phylogenetically, mtDNA sequences are related more closely to those of free-living bacteria than they are to the remainder of our DNA (which is housed in the nucleus, or command center, of each cell).

These findings helped give rise to the "endosymbiosis" theory, now widely accepted. The word "symbiosis" describes any situation in which dissimilar organisms live in close association, often to mutual benefit. The prefix "endo" comes from the Greek word for "within" and implies a cozy alliance. The idea is that the present-day genomes (genetic contents) of mitochondria descended from free-living bacteria that in ancient times (more than a billion years ago) entered into a tight collaborative relationship with primitive cells bearing precursors of many of the genes now present in our cells' nuclei. Over evolutionary time, this symbiotic relationship became so intimate that mitochondria today could no more live without "us" than we could survive without "them." They are a fully integrated part of our biological makeup.

Plant cells contain mitochondria too, as well as another type of organelle—the chloroplast—that houses genetic material also of probable endosymbiotic origin. Like mitochondria, chloroplasts are critical for cell survival and growth, in this case playing key roles in the photosynthetic processes that generate a plant's food from carbon dioxide and water. Furthermore, like mitochondria, chloroplasts contain small pieces of DNA (cpDNA) that are related more closely to some kinds of free-living bacteria than to the plant's own nuclear genes.

Overall, when viewed with respect to their microbial origins, the mitochondria and chloroplasts now living within the cells of higher organisms give special added meaning to the notion of the unity and diversity of life.

Jumping Genes: Nature's Real Movers and Shakers

Transposons are another class of microbial-like creatures with whom plants and animals have intimate associations. Also known as jumping genes or mobile elements, these tiny DNA sequences can inhabit cells in astounding numbers. For example, one type of mobile element (known as Alu) occurs in hundreds of thousands of copies in every human cell. Alu sequences arose in the primate lineage about 40 million years ago, and these tiny genetic nomads have perpetuated themselves within our cells ever since. Collectively, these and other jumping genes make up an incredible 50 percent of the human genome (our total complement of DNA).

Rather like mitochondrial DNA (discussed in the previous essay), mobile elements are an integrated component of each organism that houses them. Scientists have discovered many families of jumping genes in various plants and animals and have given them whimsical names such as gypsy, tourist, vagabond, castaway, wanderer, pioneer, pogo, Magellan, jockey, hopscotch, mariner, stowaway, and hobo. Despite their cute names, jumping genes basically are intracellular parasites with their own selfish evolutionary agenda, which is to self-perpetuate by making copies of themselves and hopping from one chromosomal site to another. From the standpoint of a transposable element, humans are nothing more than a convenient habitat suitable for producing more selfish elements.

Mobile elements seem to behave as if they have minds of their own. Most such elements are composed solely of genes that facilitate their own replication and movement, even if this comes at host expense. When a copy of a jumping gene inserts itself at a new chromosomal location, it sometimes disrupts the function of host genes adjacent to the insertion site. This can have serious health consequences. Several genetic disorders in humans, including hemophilia, leukemia, and some breast cancers, have been linked to the destabilizing influence of jumping genes.

Not surprisingly then, conflicts frequently arise in the co-evolutionary dances between jumping genes and their landlords. Host genomes may evolve mechanisms that in effect silence or jail renegade elements, but the elements apparently resent such policing activities and sometimes devise ingenious countermeasures. Occasionally, arrangements are worked out that mutually benefit the jumping gene and its host genome, as, for example, when the two parties collaborate to provide a cell with some metabolic capability that it otherwise lacked.

In regard to general physical structure, mobility, and evolutionary agendas, many jumping genes are strikingly reminiscent of a large class of infectious viruses known as retroviruses (which include the causal agent of AIDS). One important difference is that unlike jumping genes, which normally stay within a particular host organism and its descendents, retroviruses encase themselves in exploratory protein capsules that can launch forth from the original host to infect unrelated organisms. Nevertheless, the behavioral resemblance between jumping genes and certain viruses has raised several compelling questions for genetic investigation. For example, are jumping genes phylogenetically allied to retroviruses, and if so, which came first in evolutionary history?

Many jumping genes share with retroviruses a gene involved in the element's replication. Close inspection of this gene in numerous transposons and viruses suggests that these two tiny forms of life indeed are evolutionary relatives, part of the same extended family pedigree. Furthermore, the structure of that family tree implies that retroviruses probably evolved from jumping genes, rather than the converse.

Thus, jumping genes, too, evidence the great complexity of life within an individual. Each of the 100 trillion living cells that make up a human body is like a rich ecological community inhabited by tens of thousands of genes that have their own individualized evolutionary strategies and agendas.

2

Clones and Chimeras

Is cloning a horrible sin
For producing duplicate kin?
Some people bemoan
The thought of a clone;
Others don't—identical twins!

The word "clone," as used in this section, will refer to any set of
genetically identical organisms (or, when used as a verb, as the
means by which such organisms are produced). Recent breakthroughs in
genetic engineering have made it possible for scientists to clone prize
specimens of domestic animals such as sheep and cows. They also have
raised the sobering prospect of cloning human beings, a technically fea-
sible possibility that has prompted understandable concern and ethical
debate.

For Mother Nature, however, cloning is nothing new. Microbes do it routinely, every time they divide and multiply asexually. So too do many other creatures, even several species of "higher" animals, who sometimes have added bizarre twists to the process. For example, in several species of fishes, amphibians, and reptiles, females produce clonal offspring by laying eggs that develop directly into new individuals, without the benefit of fertilization by sperm. These progeny are identical in genetic composition to one another and to the cells of their one-and-only parent. This reproductive operation is known as parthenogenesis, or virgin birth. Species that procreate by this means typically consist solely of females. Various molecular-genetic approaches, collectively known as DNA fingerprinting, are suited ideally for distinguishing, in nature, clonemates (the products of asexual propagation) from non-clonemates (offspring resulting from sexual reproduction). Thus, even when an organism's reproductive mode is hidden from direct visual inspection, straightforward molecular assays of parents and their progeny can reveal the finest and most intimate details of these creatures' proliferative habits.

A chimeric individual is rather like the reverse of a clone. In other words, a "chimera" is an animal or plant specimen that is composed of a mixture of genetically different cells that have stemmed from separate fertilized eggs. DNA fingerprinting methods again are powerful for identifying chimeras, even when direct inspection provides no clue as to the composite genetic nature of such individuals.

Nature's Clone-Making Mammal In 1997 you may

have read dramatic accounts of how scientists first constructed a mammalian clone in the laboratory. Dolly, the now-famous lamb, was engineered to be genetically identical to her biological mother. In a technological coup, geneticists removed the nucleus (containing DNA) from a mammary cell of the soon-to-be genetic mother, and microsurgically transferred it into an egg cell (whose nucleus had been artificially removed) from another ewe. The engineered egg then was returned to the womb of the surrogate sheep, who several months later gave birth to lamb Dolly, a genetic clone of her biological mother.

This accomplishment truly was stunning, but did you know that many mammals naturally produce clones too, entirely without human intervention? One form of clonal reproduction begins when a fertilized egg divides a few times in the womb before initiating embryonic development. In humans, this can lead nine months later to the birth of identical (monozygotic) twins. This phenomenon, known as polyembryony, occurs sporadically in many mammal species. However, only in armadillos does polyembryony happen consistently, in each and every pregnancy.

Armadillos are cute little beasts dressed in plates of leathery armor from head to tail. Several species inhabit South and Central America, one of which (the nine-banded armadillo, *Dasypus novemcinctus*) colonized North America within the last century and is now a common sight grubbing along roadsides in the southeastern states. The first suspicion that nine-banded armadillos are polyembryonic came early in the 1900s when it was noticed that females invariably give birth to same-sex litters, typically consisting of four brothers or four sisters. However, direct confirmation that armadillo siblings are genetically identical had to await the application of DNA fingerprinting methods. The molecular data, gathered in 1996, proved conclusively that armadillo litters indeed are composed of clonemates.

How might this rare and freakish reproductive mode have arisen? At face value, regular polyembryony within each litter would seem to be a highly maladaptive evolutionary strategy. Unlike sexual reproduction, which creates genetically diverse siblings, polyembryony produces perfect copies of a single genetic type (genotype). Furthermore, because that genotype never before existed (it is a novel mixture of the genotypes of a mother and a father armadillo), it has never before been tested by natural selection. In life's reproductive raffle, it's as if each pair of armadillo parents follows a bone-headed strategy of purchasing multiple lottery tickets displaying the same random number.

One possible evolutionary explanation for polyembryony invokes the idea of "nepotism" (favoritism toward kin). Perhaps the clonemates in an armadillo litter help one another build dens, find food, or detect predators. If so, any genes responsible for polyembryony might have been favored across the generations if pronounced cooperation among such littermates led to higher survival and reproduction, on average. Alas, this hypothesis appears to be incorrect. Genetic analyses in conjunction with field investigations revealed that armadillo siblings in nature seldom remain together long after birth. Also, even when together, armadillo littermates in the wild show no indication of nepotistic behaviors.

An alternative explanation now appears more likely. At the outset of her pregnancy, an armadillo's uterus is highly constricted, having only one attachment site for an early embryo. Only later does the womb enlarge and make room for additional offspring, who then arise clonally from the original cells. Thus, polyembryony in armadillos might be a highly evolved strategy that is maintained by natural selection because it overcomes the severe initial limit on litter size that otherwise is imposed by the peculiar configuration of the mother's uterus.

Although this notion remains speculative, this story carries a deeper evolutionary lesson. Nature is not always as it appears. Despite the initially counterintuitive aspects of the armadillo's clonal reproductive mode, perhaps these little tanklike creatures aren't so evolutionarily bone-headed after all.

The Lizard That Dispensed with Sex

> Males had become mere nemeses,
> Bums to throw off the premises.
> All fathers were shunned
> When procreation
> Turned to partheno-geneses.

From an individual's point of view, sexual reproduction would seem to have several disadvantages. It takes precious time and energy to find and court a suitable partner, and the mating process can expose each participant to increased risks of predation and disease. Another severe penalty of sexual reproduction is that each parent must mix his or her genes with those of a complete stranger. In other words, by engaging in sexual as opposed to asexual reproduction, each parent dilutes his or her genetic contribution to each offspring by a whopping 50 percent. Given such high costs of sex, perhaps it is not too surprising that some species that once reproduced sexually have jettisoned the process altogether.

In a small pocket of land along the Texas-Mexico border lives a lizard species, the Laredo striped whiptail *(Cnemidophorus laredoensis)*, that has dispensed entirely with males, and indeed with the whole bothersome rigmarole of sexual reproduction. This species consists solely of females. About a foot long from head to tail and dressed in a striped outfit of brown and cream, they have evolved a unisexual reproductive mode known as parthenogenesis. In this process, a female produces "diploid" eggs, each carrying two sets of chromosomes (rather than the single set of chromosomes characteristic of "haploid" eggs in most other animal species). Without benefit of fertilization by sperm, these diploid eggs when laid develop directly into a new generation of female lizards. Thus, babies from a given female are genetically identical to one another and to their mother.

How did this parthenogenetic species arise? According to genetic evidence, the Laredo whiptail lizard originated via hybridization between two closely related species with conventional sexual reproduction: the six-lined racerunner *(C. sexlineatus)* and the Texas spotted whiptail *(C. gularis)*. Thus, the unisexual species now carries two distinct genomes (sets of DNA), one from each of the two bisexual lizard species that produced it. Both of these genomes are transmitted intact to the progeny of each unisexual female, without the shuffling of genes that occurs under normal sexual reproduction.

Actually, about a dozen all-female species of whiptail lizards occur in the American Southwest. Genetic markers have confirmed that each asexual form arose via hybridization between related sexual species, and they also have uncovered the particular parental species involved in each such cross.

In most animals and plants, one of the great advantages of sexual reproduction is that it produces genetically variable progeny within a brood or litter. Thus, at least some babies in every family stand a reasonable chance of being well adapted to environmental conditions in which they may find themselves. These profound benefits presumably often outweigh the costs of sexual reproduction mentioned above. In *Cnemidophorus* lizards, the hybrid origins of the asexual species may be crucial. By carrying a mixture of divergent genomes derived in each case from two separate species, each unisexual lizard retains an evolutionary store of genetic variation that may help to compensate for the sexual mode of reproduction it has abandoned.

Amazon Sexual Parasites
More than seventy taxonomic species of reptiles, amphibians, and fishes are made up solely of females. Remarkably, all these unisexual forms have arisen at various times in the evolutionary past from hybridization between closely related sexual species. Some of the unisexuals, like the whiptail lizards described in the previous essay, reproduce by parthenogenesis, wherein an unfertilized

egg develops directly into a new individual genetically identical to its mother. An interesting variation on this procreative theme involves "gynogenesis," which can be thought of as asexual reproduction but with just an added dash of sex.

Gynogenetic females, like parthenogenetic ones, produce eggs by a mitosis-like process that perfectly replicates the genotype (genetic makeup) of the mother's cells. Although these eggs remain unfertilized, they nonetheless can develop directly into new daughters who, therefore, are genetically identical to one another and to their respective mothers. The sexual twist comes from the fact that each unfertilized gynogenetic egg requires a little physical "poke" from a sperm cell before it can begin dividing and proliferating to form a new embryo.

Since these gynogenetic species consist of females only, where does the sperm come from? Amazingly, it comes from males of a closely related sexual species. Take, for example, a gynogenetic species of fish that lives in the arroyos of northeastern Mexico. Appropriately, this all-female species is named the Amazon molly *(Poecilia formosa).* Before each inch-long female, who bears her young internally, can reproduce, she must mate with a male from a related sexual species of *Poecilia.* This male gets no genetic reward for his amorous services, however, as each resulting daughter carries her mother's genes exclusively. Thus, the males are said to be "sexually parasitized" by the reproductively self-serving gynogenetic females.

Molecular genetic markers have contributed greatly to our scientific understanding of the Amazon molly's evolutionary origins. Although both morphological and geographical evidence strongly hinted that *P. formosa* arose from a hybrid cross between two sexual species *(P. latip-*

inna and *P. mexicana*), molecular data from various genes finally confirmed that this was indeed the case. The molecular data revealed more. Recall that mtDNA is maternally inherited in most animals. Thus, when mitochondrial genes were added to the evolutionary analyses, they showed unequivocally that the hybridization event giving rise to Amazon mollies was between a *P. mexicana* female and a *P. latipinna* male. Furthermore, the mtDNA gene sequences in Amazon mollies were remarkably similar to those of *P. mexicana* individuals alive today, indicating that the secret rendezvous in this hybrid affair took place within the last few thousand years.

In 1978 a famous evolutionary biologist, M. J. D. White, lamented that "we are never likely to know which species was the female parent" in the prehistoric hybridization events that have produced unisexual taxa. Ironically, just one year later, revolutionary mtDNA methods were introduced that permitted and soon led to the recovery of precisely such information. There's a broader message here. Just as we should never underestimate the ingenuity of nature in solving life's ecological and evolutionary challenges, so too should we be cautious not to underestimate the power of innovative science in addressing many of nature's seemingly intractable mysteries.

The 100,000-Year-Old Clone An even more peculiar

form of reproduction is displayed by a few unisexual (all-female) species that take sex one step further than was true under gynogenesis (discussed in the previous essay). Under this quasi-sexual way of procreating, known as hybridogenesis, each female again requires the carnal services of a male from a related sexual species. However, unlike gynogenesis, in this case the male makes a bona fide genetic contribution to his daughters. Each daughter inherits essentially as many genes from him as she does from her mother. Alas, the father's genetic legacy will be fleeting, totally in vain from an evolutionary point of view. The following explains why.

When a hybridogenetic female matures, she provisions her eggs solely with her mother's genes. This occurs in the female's ovaries via an aberrant cellular process that discards all the chromosomes that the female had received from her father. Thus, any of her own daughters will carry genes from their maternal grandmother, but none from their other grandparents.

As a consequence of this curious reproductive game, each sexual male again has been duped, or sexually parasitized. Generation after generation, the hybridogenetic females transmit, intact, a maternal "hemiclone" that remains unaltered genetically (except for an occasional new mutation). It is called a hemiclone because the genes are identical only to those of the "better half" of the mother (its maternal lineage). Thus, hybridogenesis overall is a strange mix of the elements of sexual and clonal reproduction.

A fine example of hybridogenetic reproduction occurs in another group of inch-long aquatic creatures—the *Poeciliopsis* fishes in the arroyos of northwestern Mexico. In these streams, each of several all-female species (for example, *P. monacha-lucida* and *P. monacha-occidentalis*) coexist with related bisexual species, sexually parasitizing their males.

How old, evolutionarily, are the hybridogenetic lineages? Molecular analyses revealed that several different hemiclones, each having arisen from a separate hybridization event, exist in the hybridogenetic sisterhood of *Poeciliopsis* fishes. Based on the genetic evidence, one of these maternal lineages was estimated to be about 100,000 years old. Thus, this all-female lineage originated in rather ancient times from an interspecific hybridization event between the sexual species *P. monacha* and *P. occidentalis*. Ever since, this remarkable strain of female fish has perpetuated itself by sexual parasitism on *P. occidentalis* males.

The ages of clonal or hemiclonal lineages are of special interest because asexual reproduction by itself usually is thought to be a dead-end evolutionary strategy for most vertebrate organisms. Lacking substantial genetic variation, lineages without full sexual reproduction supposedly seldom survive for long. If so, nonsexual lineages alive today should be confined to the recent tips rather than the ancient trunks and branches of evolutionary trees. Although the 100,000-year-old hemiclone in the *Poeciliopsis* fishes is ancient by the short standard of human lifetimes, and was record-setting at the time of its discovery, it doesn't really violate this conventional wisdom. In evolutionary time, 100,000 years is but a brief moment, an evening gone.

When Is an Individual Not an Individual?

Except in the case of armadillos (see "Nature's Clone-Making Mammal," above), seldom are the multiple embryos and fetuses that share a mammalian womb genetically identical to one another. The more common situation in humans and other species is that twins (or triplets, and so on) begin life as different fertilized eggs. These fraternal (or "dizygotic") twins are no more alike genetically than any other full-sib brothers and sisters from separate pregnancies.

Marmosets and tamarins are small primates that normally give birth to two or more fraternal offspring per pregnancy. These species live in highly social groups in tropical rainforests and are atypical among primates in the degree to which reproduction is monopolized by one female in each troop. Furthermore, adult and subadult nonbreeders, usually but not invariably earlier progeny of the primary breeding couple, routinely help this adult pair to rear additional children.

Why would individuals assist others in child-rearing? Such "alloparental care" usually is interpreted as adaptive for the helpers. For caregivers who are sons and daughters of the breeders, indirect but substantial genetic benefits might accrue when some of the genes that they share with their parents are transmitted to new brothers and sisters. For these or for unre-

lated aid-givers, reproductive assistance also could be self-serving, ultimately, if the helper
learns social or parenting
skills that are valuable to it
later in life, when it might gain a
better opportunity to breed directly.

An astonishing genetic discovery,
based on molecular markers, is
that an individual marmoset or
tamarin is a chimera, a creature con-
sisting of a mixture of two or more different genotypes. Twins
start life as separate fertilized eggs, but in the first month the tiny
embryos partially fuse inside the uterus, exchanging blood cells
and those of some other body tissues. Thus, although they are
physically separated before birth, each member of a set of twins
remains genetically part itself and part its brother or sister.

The discovery of this odd state of affairs has given rise to a
new, sophisticated mathematical argument for alloparental care
in these primates. It hinges on the suspicion that by virtue of
being a chimera, each marmoset or tamarin is actually related
more closely (genetically speaking) to its parents than to its own sex
cells (sperm or eggs). If so, then one of the two sets of genes in each
chimeric twin would, in effect, devalue that animal's personal reproductive efforts in relation to any helping behavior directed toward that
individual's parents. Thus, genetic chimerism might further tip the scales
toward helping behavior by predisposing marmosets and tamarins to
parental assistance.

This complicated and speculative hypothesis remains to be tested critically. In the meantime, what seems far more certain is that the chimeric
marmosets and tamarins give an entirely new perspective on individual
and family ties.

Chimeric Sea Squirts

Chimerism is extremely rare in backboned animals (fishes, amphibians, reptiles, birds, and mammals) but in some species of plants and small marine animals genetically distinct individuals routinely fuse into a single composite specimen. A case in point involves a squishy little marine creature appropriately named the sea squirt, *Botryllus schlosseri*. Sea squirts are tunicates, a taxonomic group that is thought to be related rather closely, in evolutionary terms, to the vertebrates (animals with backbones).

The sea squirt lives in subtidal apartment complexes, each colony consisting of hundreds of interconnected modular units called zooids. All zooids within a colony are genetically identical, having come from asexual proliferation tracing back to a single fertilized egg. However, when two separate colonies grow and come into contact, they sometimes fuse at their margins such that the zooids become jointly embedded within an organic matrix (known as a tunic, hence the group's name) that surrounds the now-chimeric colony.

Molecular markers permit scientists to monitor the fate of genetically different sets of cells following the initial contact of two sea squirt colonies. Theoretically, these two cell types might cooperate closely, or, alternatively, they might compete, parasitize one another, or even kill their opponents in a fierce scramble for space and resources.

Empirically, genetic markers have revealed which outcomes actually take place. Many colonies fail to fuse. When they do fuse initially, zooids of one genetic type sometimes are destroyed quickly by those of the other. In other cases, cells of both genetic types persist for at least several months and even may be distributed to sex cells as well as to body tissues in the colony. When this occurs, both of the genetic partners in the chimeric complex presumably can transmit genes to the next generation. All these varied outcomes are mediated in part by genetically based "kin recognition" systems. For example, favorable fusion responses take place more frequently when close genetic relatives are involved.

Other colonial marine creatures that routinely fuse or partially fuse in nature include various sponges, corals, hydrozoans (animals whose colonies look like bantam seaweeds), and bryozoans ("moss animals," who do indeed resemble moss). In most of these groups, little is firmly known about the genetic relationship between the interacting parties, or how this might relate to cooperative or abusive behaviors of the participating cells. Perhaps, as in the sea squirts, molecular markers will someday help to sort all of this out.

The Strangler Fig Gang
Strangler figs aren't your ordinary trees. For one thing, they are pollinated by tiny wasps that themselves develop within the fig tree's own flowers. Wasps of both sexes grow, pupate, and mate within these blossoms. Periodically, female wasps emerge from their floral homes by the thousands. Bearing pollen dusted onto them by the fig flower, they fly to another fig tree, where they lay their eggs and, in the process, perform the tree's pollination services.

Strangler figs also have a highly peculiar growth habit. Rather than sprouting directly in the ground, their seeds germinate in the humus-filled crotches of other tree species. Inadvertently, the seeds were deposited there by a fig-eating animal such as a monkey or a toucan bird. From that well-fertilized seed (which probably passed through a digestive tract), each fig's shoots then bolt upward and its roots downward, encircling the branches and trunk of the host plant in a not-so-loving embrace of woody latticework. The fig's roots finally reach the ground, firmly anchoring the entire structure. The ensheathed host tree eventually dies by strangulation and rots away, leaving a free-standing yet hollow strangler fig tree as its living tombstone.

Most species of fig tree are widespread in tropical forests, but individuals are distributed sparsely. Thus, from both an academic standpoint and for conservation efforts, it is important to know how many genetically different strangler fig trees inhabit a particular area. Before such tallies can be conducted, it first must be established how many different genetic individuals constitute what looks like a single fig tree. Molecular markers have provided the answer.

What appears by naked eye to be only one fig tree often proves upon molecular examination to be a chimera, composed of more than one genetic individual. This apparently happens when several fig seeds are deposited simultaneously in a particular crotch of a host tree. Later, the resulting shoots from these separate seeds grow and intermesh into a single structural entity that looks for all the world like only one tree but actually consists of a gang of distinct genotypes.

In one study site in Panama, nearly all the fig trees proved to be chimeras, consisting of two to eight, and perhaps more, genetic individuals each. Thus, the true number of fig trees is several times higher than previously supposed. This is a happy finding for conservation efforts because local population numbers of several fig tree species otherwise appeared to be perilously low. It is also a happy finding for the wasps, whose life cycle and evolutionary fate are linked so intimately to those of the strangler figs. This unanticipated genetic discovery, however, may not be such favorable news for the kinds of trees that the strangler figs suffocate.

3

Hermaphroditism

Behold the hermaphrodite
A part male, part female sight
But no one can say
If (s)he's straight or gay
'Cause they'd be part wrong, part right!

A hermaphroditic individual produces female *and* male gametes—that is, both eggs and sperm (in animals) or eggs and pollen (in plants). In some cases, these two types of sex cells are produced at the same time, and the individual is said to be a simultaneous hermaphrodite. In other cases, they are produced at different times in the life cycle. If an individual produces sperm early in life and switches to eggs later on, it is called a protandrous hermaphrodite. By contrast, a protogynous hermaphrodite produces eggs first and then sperm later. The timing of gamete production is important—if eggs and sperm are produced at different life stages, this precludes the possibility of an individual fertilizing itself.

Collectively, hermaphroditic species have sexual options ranging from exclusive self-fertilization (selfing) to exclusive outcrossing (mating with other individuals). Each form of reproduction has genetic consequences, yet which reproductive option has been employed by a given species or individual is seldom apparent from field observations alone. This is where molecular genetic markers can be of great service.

The usual approaches are to deduce the reproductive mode either through genetic parentage analyses of particular offspring, or by examining how genes are distributed within a hermaphroditic population. Selfing and outcrossing yield different genetic signatures. For example, each batch of progeny resulting from selfing carries alleles (forms of a gene) from only one parent, who in effect had mated with itself. By contrast, each batch of outcrossed progeny carries a mixture of genes from two separate parents. In the context of parentage assessment, genetic markers can help to determine whether an individual had one parent or two. At the population level, selfing is an extreme form of "inbreeding," and it can result in the appearance of highly inbred strains that are recognizable in genetic assays.

Some species show outlandish variations on the hermaphroditic theme—they contain mixtures of dual-sex and one-sex individuals. The matrimonial games in such species can get complicated, both in the day-to-day lives of individuals and in regard to the evolutionary transitions among alternative conjugal systems. Genetic markers can help to sort out this kind of sexual confusion as well.

The Fish That Mates with Itself In the muddy mangrove swamps of southern Florida and the West Indies lives a drab, inch-long killifish *(Rivulus marmoratus)* that hardly behaves like a fish at all.

Each adult lives in a tiny pocket of water deep inside the meter-long burrow of a herbivorous crab. It spends weeks or months entombed in this dark, stagnant pool and can emerge only on rare occasions when the crab's burrow is flooded by high tides or after a heavy rain. Even then the killifish's time in the sun is brief, for it soon retreats to another crab burrow, or perhaps to a narrow cavity in a mud-buried log.

What the retiring killifish lacks in size and charisma it more than makes up for by its scientific fame for an unusual reproductive lifestyle. Each individual is part male and part female. Inside each mature fish is a lump of reproductive tissue known as an ovotestis—a combination of ovary and testis—that produces eggs and sperm at the same time. Furthermore, this simultaneous hermaphrodite can self-fertilize, as suggested by the fact that individuals reared alone in aquaria may continue to produce offspring. Self-fertilization happens when the eggs and sperm unite inside each fish's abdomen. The fertilized eggs then are laid into the water where the embryos develop and hatch to begin the next generation of killifish.

Although these killifish routinely self-fertilize when reared in solitary confinement in aquaria, do they normally do so in nature? Perhaps these little fish aren't so antisocial after all. Maybe when given the chance they freely mate with other hermaphrodites and thereby mostly cross-fertilize one another. This possibility was tested critically by examining molecular genetic markers in natural populations.

It was found that self-fertilization apparently predominates in wild killifish as well. At Floridian sites where this species was examined, genetic markers revealed that populations consist of highly inbred strains. Selfing is the most extreme possible form of inbreeding. Via self-mating, generation after generation, genetic variation within each killifish strain has declined to the point where all individuals are nearly identical

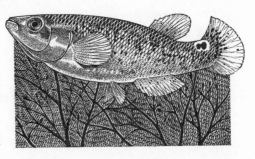

to one another. In essence, members of each strain have become clone-mates. However, different strains show pronounced genetic differences, since rarely if ever do they interbreed.

The rivulus killifish is the only species of vertebrate animal currently known in which individuals propagate themselves almost exclusively by self-fertilization. Each fish "self-mates," and each specimen has only one biological parent rather than a separate mother and father. For all practical purposes each individual is genetically identical to its one and only parent, as well as to its siblings.

Many questions remain. For example, how does this little fish manage to balance properly its male-directing and female-directing hormones so as to produce both eggs and sperm at the same time? In most other species of vertebrate animal, these hormones tend to be mutually at odds. Another question is why don't these lineages suffer from "inbreeding depression"? Inbreeding depression is a widespread biological phenomenon in which matings among close kin (such as first cousins) produce offspring with serious genetic defects. The killifish are among the most inbred creatures on Earth, yet they seem to do just fine. Perhaps in the past there was heavy mortality associated with selfing, but today we see only those fortunate killifish strains that survived the initial rounds of inbreeding depression. Once through the initial trauma, any surviving inbred strain then would have been "purged" of genetic defects.

The rivulus killifish in Florida is rare and threatened. This is a shame, because this species' unique reproductive lifestyle among vertebrate animals makes this retiring little creature a valuable scientific treasure.

How Snails Sow Their Oats If you've traveled through
central California, you might associate the foothills with their most conspicuous and characteristic vegetation, the slender wild oat (*Avena barbata*). For most of the year, great stands of this grain form rolling waves that bathe the hillsides in a golden brown, converting to a vibrant green after spring rains. The slender wild oat is native to the Mediterranean re-

gion, from whence it was introduced to California by Spanish missionaries some 400 years ago. It is a hermaphroditic species. Individual plants, which produce eggs and pollen simultaneously, sometimes self-fertilize and sometimes cross-pollinate one another.

Another hermaphroditic species introduced from the Mediteranean region into the New World is the tiny land snail *Rumina decollata*. This inch-long animal has a slender glossy shell sculptured with fine spiral lines. It lives in gardens and parks, hiding under stones or in forest debris. As in the slender wild oat, these snails routinely reproduce by selfing, but they also can mate with and thereby cross-fertilize other individuals. Undoubtedly, selfing has contributed to the success of both the wild oat and the snail as colonizers, because a single individual carries both male and female gametes and by itself can start a new population.

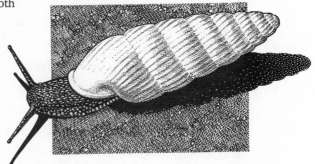

In addition to sharing an ancestral homeland, being good colonizers, and displaying a hermaphroditic lifestyle, what else do the slender wild oats and the slender land snails have in common? Researchers showed that these species are remarkably alike in how their populations are structured genetically. The "genetic structure" of any species refers to how its genes are arranged in space across the landscape. In both the snail and the wild oat, this organization is striking, and we now know why.

Species that can reproduce both via selfing and outcrossing are said to have a "mixed-mating system," and in some ecological respects they have the best of two worlds. Over the generations, selfing inevitably leads to highly inbred strains, within each of which all individuals are nearly identical genetically. Thus, any selfing strain well adapted to local conditions is at an advantage in the short term because it continually produces offspring with identical copies of a field-tested genotype. However, when environmental conditions change, outcrossing is likely to be advantageous because parents then produce genetically variable offspring, some of whom might fare better in the altered regime. Thus, in theory, a mixed mating system could lead to highly adapted strains, each with a peculiar combination of genes well suited to the local environment.

This theory has been tested by using molecular markers to assess the genetic structure of both oat and snail populations. In wild oats studied in California, self-pollination proved to be common at most locales. Over time it has led to the appearance of two distinctive genetic strains, one (the xeric type) characteristic of semiarid savannas bordering the Sacramento-San Joaquin Valley, and the other (the mesic type) dominating wetter soils of the coastal ranges and foothills. Plants with scrambled genotypes were observed only in soils of intermediate wetness. These were the progeny of cross-pollination events between the xeric and mesic types.

Likewise, in snail populations studied in southern France, selfing proved to be common at most locales, and over time it, too, has produced two distinctive genetic strains, one (the mesic type) usually associated with wet conditions under rocks and logs, and the other (the xeric type) occupying dryer habitats. Occasional outcrossing also was documented by the discovery of individuals with mixtures of genes from the mesic and xeric strains. The similarity of population-genetic patterns in the snails and wild oats is an outstanding example of "evolutionary convergence," a phenomenon wherein similar outcomes (in this case of the mating system and resulting population structures) are observed in very different organisms.

The clear association between ecological regime and genetic makeup in both the wild oat and land snail demonstrates the power of natural selection operating within the context of a mixed-mating system. Via occasional outcrossing, each species continually generates new genetic combinations for scrutiny by natural selection, and some of these prove to be highly adapted to local ecological conditions. Then, via the intense inbreeding that accompanies a return to selfing, such genetic combinations are perpetuated more or less intact, generation after generation.

These are the processes that profoundly shape the population genetic structures of hermaphroditic species with mixed-mating capabilities. From the vast number of wild oat and snail genotypes imaginable, only two in each case now predominate across California and France, respectively.

Barriers to Self-Pollination

For many animal and plant species, incestuous matings can be a bad thing. That is to say, inbred progeny (those resulting from matings among close relatives) frequently show poor survival and reproduction compared with their outbred counterparts. There are two main aspects to inbreeding depression. First, inbreeding brings together identical alleles (forms of a gene), some of which harm the individual when present in double-dose (in homozygous condition). Second, lacking substantial genetic variation, an inbred strain may respond poorly to environments that change over time or vary in space.

Fertilizing oneself, or selfing, is the most incestuous form of inbreeding. Yet, it is a real possibility in the large fraction of the botanical world that consists of hermaphroditic plant species that produce both eggs and pollen at the same time. Botanists refer to many of these hermaphroditic plant species as "monoecious," as opposed to "dioecious" forms, in which each individual produces either eggs or pollen only (and cross-fertilization therefore is obligatory). Although hermaphroditism permits selfing, this option is not necessarily exercised. Thus, an important scientific question is as follows: How regularly do hermaphroditic plants forgo selfing and thereby avoid the potential pitfalls of inbreeding?

Based on genetic parentage analyses and other lines of evidence on reproductive modes, the rates of selfing versus outcrossing have been examined for more than 200 hermaphroditic plant species. The results show that roughly 50 percent of these species, ranging from many wind-pollinated trees to such familiar roadside weeds as Europe's common poppy *(Papaver rhoeas)*, avoid self-fertilization rather scrupulously. Instead, when they reproduce sexually, they primarily outcross with other individuals.

The immediate mechanisms by which these monoecious plants avoid selfing are varied. In some species, male and female flowers of an individual mature at different times, or at different locations on the plant. In other species, male reproductive organs (stamens) and female parts (stigmas) occupy the same flowers but are positioned such that the mechanical transfer of pollen is unlikely. But for many plant species, Mother Nature has devised an even more ingenious ban on reproductive self-gratification: "self-incompatibility genes."

These genes work as follows. When a stigma and a pollen grain share an identical allele at a self-incompatibility gene, the female reproductive tissue actively discriminates against the pollination process. For example, the stigma may inhibit the germination or growth of the tiny pollen tubes that otherwise transfer male sex cells to the precious eggs. It turns out that self-incompatibility genes are among the most variable genes known in plants. Thus, in any outcross event, the female stigma and male pollen are likely to be genetically different at the self-incompatibility gene, such that there is no barrier to the fertilization process. In most cases, only when the stigma and the pollen grain come from the same parent (in other words, when there is a potential self-fertilization event) do these cells normally share alleles at the self-incompatibility gene. Then, the pollination barrier kicks in, nipping the fertilization process in the bud.

By virtue of self-incompatibility genes, the female side of an individual plant in effect ties the reproductive tubes of her own male half, thereby rejecting his untoward amorous advances. Such is the botanical world's natural method of birth control from matings among the closest of kin.

More Sexual Confusion

Part male, part female creatures (hermaphrodites) are of common occurrence in many invertebrate animals, and hermaphroditism is *the* most common breeding system in flowering plants. However, this form of reproduction is not the most jumbled of sexual lifestyles to be encountered in nature. To add to the confusion, some species are part hermaphrodite, part not. In many plant species, for example, some individuals are hermaphrodites and others are females only, a sexual system referred to as gynodioecy. The reverse of gynodioecy is androdioecy, wherein some individuals are hermaphrodites and others are strictly males.

Androdioecy is especially puzzling to some mathematically inclined evolutionary biologists, who have concluded that this breeding system could not possibly be maintained in nature for long. Their mathematical arguments against androdioecy do not arise from any moral opposition to this bizarre carnal set-up, but rather from the difficulty that natural selection would have in permitting the persistence of male individuals in a population otherwise consisting solely of hermaphrodites.

The problem in accounting for the continuance of androdioecy is that, in theory, pure males would have to fertilize, on average, twice as many eggs as would a typical hermaphrodite (whose genes are passed through both eggs and sperm). This would seem to be a nearly impossible task for such males. The hermaphrodite has the advantage because it has two available routes to a genetic payoff (eggs and sperm), compared with only the sperm route for pure males. Androdioecy is indeed extremely rare in the natural world, as evolutionary theory generally predicts. However, the phenomenon *is* firmly documented in the Durango root *(Datisca glomerata)*, a perennial herb that occupies stream habitats from northern California to the Baja Peninsula.

How did androdioecy evolve in *D. glomerata*? There were two competing hypotheses: that androdioecy is a half-way stage in the evolution of dioecy (separate sexes) from hermaphroditism; and that androdioecy is a mid-step in the evolution of hermaphroditism from dioecy. Both hypotheses interpreted androdioecy as an interim system between hermaphroditism and separate sexes, but they were exact opposites with respect to the predicted evolutionary direction.

The challenge in the ensuing genetic detective work was to determine whether the immediate ancestors of *D. glomerata* were hermaphroditic or dioecious. By examining suitable DNA sequences, researchers were able to deduce the phylogenetic placement of *D. glomerata* in an evolutionary tree that included several related species with known reproductive patterns. By mapping the breeding systems onto this extended family tree, scientists concluded that the immediate ancestor of *D. glomerata* was most likely a dioecious species, rather than a hermaphrodite. Thus, *D. glomerata* has been caught in an evolutionary transition from having two separate sexes to someday perhaps consisting only of hermaphroditic individuals.

In 1877 Charles Darwin published a book on the varied reproductive systems of plants in which he speculated on the evolutionary intermediacy of andro- and gynodioecious species to hermaphroditic and separate-sex forms. If Darwin came back today, no doubt he would be pleased and amazed to see how genetic analyses have revealed the likely details of such evolutionary transitions.

The Fish That Changes Its Sex
Have you ever felt a desire to experience the world through the mind and body of the opposite sex? Actually, some species in nature must do precisely that as a normal part of the life cycle. These are the protandrous and protogynous hermaphrodites, who begin life as males or females, respectively, and then at some point switch over to become a member of the opposite gender. The transition is complete in all respects: behavior,

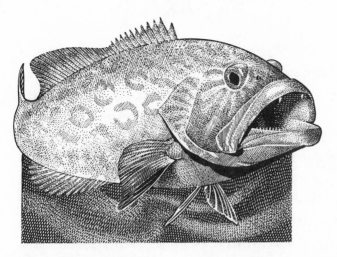

physiology, and function. In some species, what had been a male becomes a female, and in other species, what had been a female becomes a male.

Take, for example, the gag grouper *(Mycteroperca microlepis)*, a popular reef fish avidly sought by commercial and recreational fishermen in the southeastern United States. Individuals in this species actually do begin their reproductive careers as egg-laying females, but many of them later in life change into sperm-producing males. Thus, males in the species tend to be older and larger than females. Externally, these females-turned-males are recognizable by their dark copper-colored stomachs, and they also tend to be far more aggressive and territorial.

Relatively little is known about the life history of gag groupers, but it is suspected that individuals return year after year to the same aggregation site (perhaps their natal home) to spawn. These schools of fish, typically at 150- to 350-foot depths in the ocean, are easy targets for hook-and-line fishermen anchored over the proper reef. The ease with which gag groupers can be caught has raised important questions on how to manage this species and protect it from extinction. Several issues that can be addressed by genetic markers are relevant to the discussions, as are (surprisingly) the gag groupers' sex-change operations themselves.

Molecular assays of DNA-level polymorphisms have revealed that aggregations of gag groupers separated by as little as 100 miles sometimes show significant genetic differences. This finding suggests that genetic exchange among groupers on different reefs is limited, and that populations have "drifted" apart somewhat in their genetic makeup. "Genetic drift" is a phrase used by biologists to describe random genetic fluctuations in populations that are spatially separated. The effects of genetic drift are more pronounced in small than in large populations, and in those that have limited as opposed to extensive genetic contact with one another (via mating and dispersal). Thus, the genetic evidence suggests that gag grouper populations on the Atlantic seaboard are smaller and more isolated than formerly supposed.

One important factor that might contribute to the rather small size of particular gag populations has to do with the skewed "sex ratios" (numbers of females versus males) stemming from the fish's sex-changing lifestyle. On the spawning reef, males are far more aggressive feeders than females, making them especially vulnerable to baited hooks. As a consequence, males are caught disproportionately and tend to become rare. For example, at one major spawning site off the east coast of Florida (near West Palm Beach, where fishing pressure is intense), the proportion of males has plummeted in the past two decades: female-to-male sex ratios are now 30:1, compared with a mere 8:1 before the fishing explosion.

In any sexually reproducing species, an obvious truism is that every individual has a father and a mother, regardless of how many adult males and females are present in the breeding population. In other words, no matter what the population's sex ratio, exactly one-half of the genes in any generation come from fathers and one-half from mothers. This means that a dramatic decline in the numbers of even one sex can impose a "bottleneck" on a breeding population, thereby increasing the effects of genetic drift and seriously diminishing the level of variation in the population's gene pool. This may well be what has happened to gag groupers as a result of overfishing in this century. All this has put the gag grouper in rather serious straits, and some professional fishery biologists have called for a ban on local harvests until this species can recover its more natural balance of females and males.

4

Sex, Pregnancy, and Making Babies

Many of nature's wildest behaviors concern matters of sex and reproduction. This part, and the two that follow, provide examples of unusual, comical, or downright outlandish ways that sexual organisms go about their procreative business. Much of this knowledge has come through detailed molecular genetic analyses of paternity and maternity in offspring cohorts (broods or clutches).

Mating behaviors in nature can be highly devious because they reflect resolutions of conflict within and between the two sexes over reproductive turf. Although sexual reproduction normally requires a collaboration between partners, each individual has selfish reproductive motives. Take, for example, the common situation wherein females become pregnant or otherwise expend high energy in raising babies. In such species, a male might improve his own genetic fitness (the number of offspring he produces) by mating with several females, but the fitness of his mate might be harmed by this tactic if it diminishes his investment in her offspring. Thus, a female might profit from finding a male who remains faithful to her and her children (whether or not he is the actual sire). That male, then, must be certain that he truly is the sire, or else he may invest his time and energies in rearing foster children sired by a competitor. Clearly, reproductive incentives for surreptitious matings or coercive behaviors (by both sexes) abound even in this simple situation, and what's best for the genetic fitness of the goose isn't always what's best for the genetic fitness of the gander.

Imagine now that males became pregnant instead. How might this alter the preferred reproductive strategies of males and females? Actually, males do get pregnant in some species, and genetic markers have been employed to examine what this change in the reproductive game means for these creatures' mating behaviors, and for the "sexual selection" pressures at play. Sexual selection is a form of natural selection that acts via the mating preferences of males and females, and it often leads to behavioral and morphological differences between the sexes ("sexual dimorphism").

Current ecological conditions as well as historical genetic constraints have added many further rules to the mating guidebook. The bottom line is that genetic fitness is the yardstick by which Mother Nature evaluates and compares the various reproductive strategies of individuals within a species. Plants and animals pursue their reproductive goals with great vigor and diverse behavioral tactics.

This part will begin to delve into nature's reproductive treasure chest by highlighting some interesting tales of female and male pregnancy, marriage and divorce, parental choice in offspring gender, and varied mechanisms of sex determination.

The Onus of Pregnancy

Many women readers might entertain (and bemoan) the thought that internal pregnancy is a millstone borne uniquely by female mammals. Indeed, most female birds, reptiles, amphibians, fishes, and invertebrate animals lay their eggs outside the body, thereby freeing themselves from having to haul around developing embryos and fetuses. But there are exceptions. Among the fishes, in one taxonomic family (Poeciliidae) that includes hundreds of species, females always carry their young internally and, like mammals, give birth to live babies.

One example is the mosquitofish, *Gambusia holbrooki,* whose common name comes from the species' habit of devouring aquatic mosquito larvae. As in other poeciliids, male mosquitofish have a gonopodium that really is a modified anal fin but that looks and functions like a penis. Adult males are small, and a cadre of them often follows a larger adult female like a flotilla. Seemingly never discouraged, these males behave like impertinent, zealous swordsmen, each insistently thrusting his gonopodium toward the genital opening of the female.

Lady mosquitofish parry most of these stabs simply by turning away. However, some sexual unions do, of course, take place. Then, sperm transferred by the male's gonopodium may fertilize some or all of the female's batch of about thirty eggs. As in mammals, the mother's burden is more than weight and bloating alone—a pregnant mosquitofish also provides her developing embryos with nutritional support. Internal gestation lasts about three to four weeks, after which the babies pop out to begin a free-swimming existence.

How many males typically father a mosquitofish brood? This is a classic question of paternity analysis for which molecular markers are apropos. Paternity analyses in this and other cases of female pregnancy are relatively

easy because the mother of each brood is evident and, thus, the father's genetic contribution to each embryo can be deduced readily by subtraction. For the mosquitofish populations genetically examined in South Carolina, most of the pregnant females proved to carry embryos sired by at least two different fathers. This means that each brood consisted of a mixture of full-sib and half-sib brothers and sisters, all sharing the same mother but many having different fathers. Thus, each of these females was multiply inseminated, and her brood displayed multiple paternity.

Molecular markers likewise have revealed multiple paternity in the individual clutches of a wide variety of species with internal fertilization. These range from various plants to numerous species of insects, to birds and mammals.

Multiple paternity has given rise to the now-popular notion that sperm (or pollen) from different males may compete vigorously within a female's reproductive tract for fertilization success. In the Woody Allen movie *Everything You Always Wanted to Know About Sex but Were Afraid to Ask,* one scene portrayed sperm cells as a plane-load of excited paratroopers summoning the courage to parachute into foreign female territory. Well, science can be as strange as fiction. It really is true that the millions of sperm deposited inside a female's reproductive tract must compete avidly for access to the relatively few available eggs.

"Sperm competition" has become a hot research topic in recent years because it gives recognition to a potentially powerful yet formerly underappreciated selective force in the evolutionary process. Natural selection operates not only at the individual and population levels, but also at the level of sperm competing for fertilization success within a female's reproductive tract.

Pseudo-Nuptial Flights
in Pseudoscorpions Pseudoscorpions, distant relatives of spiders, are another group of creatures with a modified form of female pregnancy. The reproductive process typically begins when a male grasps a female and deposits a stalked packet of sperm (spermatophore) onto a stick

or other solid surface. With her apparent compliance, the male then maneuvers his mate's genital opening into position over the packet, and sperm are transferred. Females mate actively. In short succession, they may receive sperm from several males and then store that sperm internally, in viable condition, for at least several days. Pseudoscorpions give live birth, and a female then carries the several dozen developing embryos in an external brood sac.

One species of pseudoscorpion, *Cordylochernes scorpioides,* native to the rainforests of Central and South America, has a remarkable life cycle that includes an association with the much larger harlequin beetle, *Acrocinus longimanus.* The flightless pseudoscorpion and its flight-capable host live in scattered old decaying trees. Often, pseudoscorpions are marooned on a single rotting tree for their entire lives, but occasionally they hitch free rides on a harlequin beetle to another tree. They do this by hiding under the wing covers of their larger mobile host, like comfy passengers on a jumbo jet.

Each cozy beetle "armpit" is an exclusive mating territory where one prime pseudoscorpion male courts any female who happens to have caught the same flight. These males also bully smaller, weaker males, many of whom never depart the tree on which they were born. Thus, each frequent-flyer male would seem to have a great mating advantage over his grounded compatriots.

However, researchers have discovered that the flying females sometimes already carry sperm that they acquired from matings within a de-

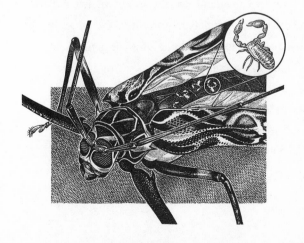

caying tree. So, do the flying males really father most of the pseudoscorpion offspring after all, or do some of the stay-at-home males also have a good chance to reproduce? As gauged by genetic data on paternity, most of the maternal broods in this species proved to include embryos from multiple fathers. Thus, any beetle-riding male can at best have sired only some modest fraction of his mate's children.

Why would a female pseudoscorpion mate with sedentary males in addition to her traveling companion on the beetle express? In theory, by copulating with multiple males, a female (in any species) might increase her genetic fitness in any of several ways. Perhaps multiple mating is a fertilization insurance policy. No matter how attractive a male may otherwise be, there is always some chance that he is completely or partially sterile. Multiple mating by a female probably lowers the risk her eggs will go unfertilized, and this might be very important for a traveling pseudoscorpion female who later finds herself colonizing an unoccupied rotting tree.

Another potential advantage to multiple mating is that a female might thereby acquire better genes for her children, on average. Suppose, for example, that after mating once or twice, a female finds a higher-quality male. Especially if she can somehow influence which of the males' sperm will fertilize her eggs, multiple mating will provide her with additional potential sires from which to choose. Finally, by mating with multiple males, a female can produce offspring with a broader collective genetic diversity, and this may be beneficial in regard to their mean survival, especially in unpredictable environments. For a roving species such as the pseudoscorpion, where pregnant individuals forever are flying off to new habitats, this too might be extremely important.

Detailed experiments have indicated that some of these benefits do indeed apply to the pseudoscorpion females. Thus, by smuggling on board some sperm from earlier copulations, a pseudoscorpion female on her nuptial flight avoids depending on a single male flying companion to fertilize her eggs. Furthermore, although some of the smaller, grounded males are barred from beetle flights, their gametes and progeny nonetheless can wing their way around.

Male Pregnancy

Mother Nature sometimes makes males pregnant too. This happens in all the 200 or so species of pipefish and seahorses, a taxonomic group (family Syngnathidae) of little animals that usually inhabit warm shallow seas. As their names imply, pipefish resemble pipe stems, and seahorses look like tiny piscine ponies. But what makes these creatures even more intriguing to marine biologists is their reproductive behavior. During mating, a female deposits her eggs into a brood pouch on the male's abdominal surface. The male then fertilizes the eggs by releasing sperm into this incubation sack. The father provides the resulting embryos with nutrients and protection for several weeks before giving birth to 30 to 1,000 babies, depending on the species. These offspring are diminutive versions of their parents and after birth are strictly on their own.

Male pregnancy, a phenomenon foreign to mammals, affords scientists a totally novel perspective on mating and reproductive behaviors. We may be accustomed to think of females as the limiting resource in reproduction, yet in pipefish and seahorses it is often the males (and their finite brood space) that limit the number of babies. In such species, sexual selection on females (via competition for mates) actually could be more intense than that on males. As a consequence, females might tend to evolve bright colors or other body ornaments to attract the sexual favors of the valuable males. Indeed, in many pipefish species, females are more festooned than males, displaying lovely stripes on their flanks, for example. This is the reverse of the usual situation in most birds and mammals, where males tend to be under stronger sex-

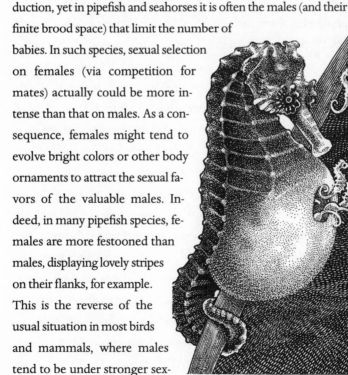

ual selection than females and often have evolved greater sexual adornments (think of a bull elk's rack, or a peacock's showy tail).

In theory, the keenness of female competition for mates (that is, the intensity of sexual selection on females) also should relate to the genetic mating system of each pipefish and seahorse species. In particular, it might be most intense under polyandry, a mating system in which some females have multiple mates (but each male has only one). Then, because some females are going to be far bigger winners of mates than others, sexual selection will most strongly favor any females who are able to catch the eye of the choosy males.

What have genetic markers revealed about the mating patterns of pipefish and seahorse species, and how might this relate to sexual selection and sexual dimorphism? In some pipefish species, each male's brood pouch proved upon genetic examination to contain embryos from only one mother, and in some cases that female had distributed her clutch of eggs between two or more males. In other pipefish species, the genetic markers revealed that a female often laid her eggs in two or more males' brood pouches in short succession, and most of these males also typically carried embryos from multiple females. In other words, both sexes had multiple mates. In seahorse species examined, monogamy was the rule (each male and female had only one mate). These genetically deduced mating patterns proved to be consistent with traditional sexual selection theory. Thus, sexual dimorphism (with females brighter than males) was greatest in the pipefish species, where females had multiple mates, and was nearly absent in the monogamous seahorses.

Another interesting finding emerged from genetic analyses of the seahorses. From field observations, most seahorse couples seem highly devoted to one another. A mated male and female greet each other daily with ritualized behaviors that include head nods, and hugs with their prehensile tails. However, genetic markers proved that babies in successive broods of a given male frequently had different mothers. Evidently, many seahorse couples had "divorced" and then "remarried." Although each pregnancy or breeding event involved monogamy, individuals commonly switched partners during a breeding season.

In traditional Chinese medicine, seahorse potions are considered an aphrodisiac and a cure for human male impotence, a folklore that stems from the supposed lifelong devotion of a virile seahorse stallion to his sole mate. Seahorses around the world are harvested and killed by the millions to support a pseudomedical trade, centered in Hong Kong, that threatens the very existence of these marvelous fish. Perhaps if consumers realized that seahorse males are not such faithful mates after all, the superstition that underlies this sad industry would disappear. Male pregnancy is a rare spectacle in the biological world, so the species that display this curious behavior are uniquely precious for the evolutionary insights they provide.

A Bird That Chooses the Sex of Its Children

Human parents sometimes wish they could choose the sex of a child. In principle, it should be a simple matter. If you want a son, just pick a sperm cell bearing a Y-chromosome to fertilize the egg. If you want a daughter, merely pick instead a sperm cell bearing an X-chromosome. In practice, of course, such parental choice of an offspring's gender is not so simple. In most mammals, humans included, whether a boy or girl is conceived during fertilization is pretty much the luck of the draw from the pool of millions of sperm that a father deposited in the mother's reproductive tract during a mating event.

Like mammals, birds have a chromosome-based sex determination system. However, in this case it's the draw from the mother's ovarian pool of eggs that determines the baby's gender. If that egg cell happens to carry a Z-chromosome, it's a boy; if it carries a W-chromosome, it's a girl. As in humans, this gametic draw in birds is more or less random. At least, that's what biologists used to think.

The Seychelles warbler, *Acrocephalus sechellensis,* is a small insectivorous species that clings to a precarious existence in its isolated home on the Seychelles Islands in the western Indian Ocean. Only a few hundred individuals survive today. Such a rare species would seem to be an unlikely candidate to give the science of reproductive biology a big surprise, but that is precisely what it has done.

Each breeding pair of Seychelles warblers occupies the same territory for as long as nine years, making only one clutch per year, each with merely a single egg. A daughter (but seldom a son) often remains with its parents for two to three years, purportedly helping to raise additional offspring. Indeed, field studies have shown that these helpers significantly increase the total number of surviving babies from parents on high-quality territories. For parents on poor-quality habitats, however, the number of children reared in future clutches actually diminishes in the presence of earlier daughters. This likely happens because under these poorer conditions, the stay-at-home daughters compete significantly with other family members for limited food.

Due to this species' endangered status, conservation biologists have had a keen interest in the birds' basic reproductive biology. One immediate challenge was to identify the sex of each baby in a nest. Adults can be distinguished by appearance and behavior, but how could the sex of a nestling be identified? In this species as in most other birds, the sex organs (ovaries or testes) are internal and in any event are not well developed in hatchlings. Thus, genetic markers were put to the identification task. By screening a drop of blood from each baby for W-chromosome genes (which only females possess), biologists were able to identify the sex of each young warbler.

That new capability led in turn to the truly big surprise. By sexing the baby birds in many nests, scientists discovered that warblers on poor-quality territories produce mostly sons, whereas those on high-quality territories generate mostly daughters. Thus, parents apparently adjust the sex ratio of their offspring in ways that appear to be highly adaptive. When daughters likely will enhance the family's genetic fitness by genuinely helping to raise more offspring from the nest, daughters are produced; but when these homebody children would be a net burden to the family's genetic fitness, mothers and fathers chiefly beget sons instead. Most of these male progeny leave the home territory and therefore don't compete directly with their parents or future siblings for food or other resources.

Exactly how these warbler couples choose the sex of their babies remains a mystery for now. In any event, let us hope that this ingenious adaptive tactic will help these highly endangered birds survive and procreate long into the future.

Roly-Poly Sex Ratios

Tiny roly-polies, or pill bugs, are familiar creatures seen by the dozens under backyard rocks or logs. Also known as isopods or wood lice, a roly-poly gets its name from its habit of curling up into a ball, for protection, when disturbed. Children and gardeners sometimes can't resist rolling these little animals about like so many marbles.

A European species, *Armadillidium vulgare,* has been of special interest to biologists for another reason: in many populations, females outnumber males by more than ten to one. After copulation, a female roly-poly stores sperm and can produce successive broods without re-mating. Mothers carry the developing babies in a special incubation pouch, and many broods consist almost exclusively of daughters.

Sex ratio in any population is the

number of females relative to males. In humans and most other mammals, an X-Y chromosomal system helps to maintain the sex ratio near 1:1 (or 50:50). However, many other sexual species depart a great deal from equal numbers of males and females. The European roly-poly is merely a clear example, and one for which a surprising explanation has been uncovered. Why do these roly-polies have such strongly female-biased sex ratios?

From recent genetic evidence, roly-poly broods that display the greatest excesses of daughters come from mothers who were infected by bacteria in the genus *Wolbachia*. These microorganisms inhabit the cell cytoplasms of roly-polies (and many other arthropod species) and are passed primarily from infected mothers to daughters. Indeed, this is the bacterium's main method of procreation.

Recall that during a fertilization event in most species, the cytoplasm of a zygote (fertilized egg) comes from the female's unfertilized egg, rather than from a male's fertilizing sperm. This means that any microbe inhabiting the cytoplasm of cells is transmitted through the maternal lineage of its host (just as is mtDNA). In other words, both sons and daughters inherit the contents of a cell's cytoplasm from mothers, but only daughters transmit cytoplasmic genes to their children. For a selfish *Wolbachia* bacterium, interested simply in the perpetuation of its kind, this transmission pattern makes a world of difference in regard to preferred evolutionary strategy.

If a *Wolbachia* bacterium happens to find itself housed in a male roly-poly, or in a female roly-poly who fails to produce daughters, it has reached the end of its reproductive line. Thus, from the microbe's point of view, it is highly beneficial to bias somehow the sex ratio of its host toward daughters. The *Wolbachia* bacterium apparently accomplishes this feat by feminizing roly-poly males, turning them during development into functional females instead. The roly-poly sex ratio in a brood may start out near 1:1, but then at least some of the male roly-polies become transfigured to egg-producing females. Through this underhanded maneuver, the *Wolbachia* microbe greatly enhances the likelihood that it will be passed on to the next roly-poly generation.

Ironically, if a *Wolbachia* strain ever were to become 100 percent successful in transforming roly-poly males to females, it might cause its own extinction (along with that of its roly-poly host). Roly-polies are sexual creatures, requiring both sexes to reproduce. Thus, if male roly-polies were eliminated entirely, both a local roly-poly population and the *Wolbachia* bacteria would meet their demise. This already may have happened on many occasions. After all, the evolutionary process has no foresight. What might seem to be a smart strategy for a *Wolbachia* strain in the short term might not be so smart if taken too far.

Wolbachia is just one on a growing list of recently discovered microbes and other selfish genetic elements, usually housed in the cell cytoplasm, that alter their hosts' sex ratios in ways that favor the microbe's near-term reproductive success. This behavior even may come at host expense. Thus, creatures like *Wolbachia* can be thought of as reproductive parasites, or even as sexually transmitted diseases.

The Social Equality of Gulls

In about one-half of the world's bird species, adult males and females are indistinguishable in appearance to the human eye. Furthermore, seldom is it possible for researchers to sex avian nestlings or "teenagers" by visual inspection alone. In part, this is because a bird's sex organs are internal rather than external. Thus, to be certain of a bird's gender, biologists in the past were forced to use invasive or even lethal dissection methods to peer into a bird's body, to view the hidden testes or ovaries directly.

In addition to the ethical concerns, there are several reasons why biologists might wish to know a bird's sex without harming the animal. For example, such knowledge is indispensable in designing captive breeding programs for endangered species, or in censusing threatened natural populations for potential numbers of breeding pairs. Biologists also are interested in how a population's sex ratio might change through time (for example, from differential mortality of males and females), or, in species with gregarious or communal behaviors, which of the two genders comprises most of the social groups.

A case of the latter point involves the brown skua *(Catharacta lonnbergi)*, a big dark seagull of polar regions in the Southern Hemisphere. In a large breeding population on the Chatham Islands (about 500 miles east of New Zealand), skuas form breeding groups composed of several territorial adults who jointly defend a nest and contribute to incubating the eggs and rearing the young. In addition, nonbreeders (youngsters and the elderly) routinely associate in social collectives known as clubs.

To all external plumage appearances, male and female skuas look alike. Thus, until recently scientists had no easy way to answer the following kinds of questions: What are the sex ratios in the skua's territorial breeding groups, and in their chicks? Do the assemblages of nonbreeders constitute all males or all females, or are they fully integrated with respect to gender? No account of the natural history of brown skuas would be complete without such basic information on the sexual composition of the various social groups.

It turns out that the two sexes *can* be distinguished readily in modern genetic assays. The trick was to identify DNA fragments confined to the W-chromosome, which is female-specific in birds (see "A Bird That Chooses the Sex of Its Children," above). A tiny drop of blood was taken from each skua and assayed for the presence or absence of a W-specific molecular marker.

The genetic assays revealed that male and female skua chicks were about equally frequent in the nests and that each breeding group was composed of one female and varying numbers of males. As for the nonbreeders, both males and females were present in particular social groups, albeit with a possible bias in favor of males. Apart from birds, there are many other vertebrate animals (ranging from tiny fishes to some of the largest whales) in which the sex of an individual is not readily apparent from morphology or routine behavior. But with the identification and use of sex-specific molecular markers, it is now quite possible to determine the sex of particular specimens noninvasively.

Extreme Social Behavior and Gender Control

Reproductively speaking, bees, ants, and wasps (the hymenopteran insects) are among the world's most amazing creatures. Take, for example, the process of sex determination. Males develop from unfertilized eggs and, thus, are "haploid" for a single set of chromosomes inherited from their queen mother. Each female develops from a fertilized egg and, thus, carries one set of chromosomes from the queen and one from her father (that is, she is "diploid"). Some of the females develop into sexual adults, fly away to mate, found a new colony, and thereby complete the cycle. Most of the females, however, are barren and serve as industrious workers in the hive or nest. These sterile housekeepers never have children of their own; instead, they raise more of their queen's offspring.

"Eusociality" is the name given to this phenomenon wherein sterile workers defend the communal colony, tend the hive or nest, and rear the queen's eggs and young. At face value, sterility would appear to be a most extreme form of reproductive "altruism" because the workers completely forgo their own reproduction and instead labor in behalf of the social collective. How could such self-sacrificial behaviors have evolved?

An intriguing hypothesis was put forth many years ago by Bill Hamilton, a mathematical population biologist. He started with the observation that eusociality and "haplo-diploid sex determination" almost always go hand in hand in

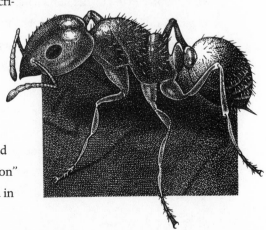

the animal world. Hamilton's insight was that this odd form of sex control alters the genetic relationships of colony members in such a way that females actually share more genes (75 percent of them) with their full-sib sisters (those sharing the same parents) than they do with their mother (50 percent). This differs from the situation in most other species, where daughters are as close genetically to their mothers as they are to their full sisters.

Thus, Hamilton argued, females under a haplo-diploid genetic setup have evolved behavioral dispositions to help raise sisters rather than produce their own daughters. By so doing, females who sacrifice their own personal reproduction to help raise some fertile sisters actually could leave more copies of their own genes to the next generation. Over time, under the influence of this type of "kin selection," genes for extreme helper behavior would spread in a population and lead to eusociality.

However, Hamilton's mathematical argument assumes that females within a colony of bees, ants, or wasps are full sibs. But, as field observations hinted and genetic markers have proved, this isn't always the case. Here's why.

When a fertile princess hymenopteran leaves her home hive to establish a new colony and become its queen, she first must mate and thereby acquire a long-lasting supply of sperm to fertilize the eggs that she will lay across her lifetime. For many hymenopteran species, molecular genetic markers have revealed that the workers within a colony can have many different fathers, meaning that their queen had mated successfully with multiple males on her nuptial flight (see "Mating Champions of the Insect World," in Part II, on page 177). Furthermore, in some species such as fire ants, the genetic data confirm that two or more queens may found a colony jointly. Thus, whether through multiple paternity or multiple maternity, or both, workers in a colony are not always full sibs. Instead, they may be either half-sibs (sharing a mother but having different fathers) or unrelated individuals (sharing neither parent). To some extent, these discoveries raise a conundrum for Hamilton's hypothesis, because a worker collaborating with half-sisters or more distant kin, rather than full sisters, presumably would do far better raising her own children.

Still, these findings on the genetic composition of many hymenopteran colonies are far from fatal to Hamilton's scenario. Quite likely, eusociality first evolved under kin selection as Hamilton envisioned, in an ancestral haplo-diploid species whose colonial workers were full sibs. Once established in evolution, eusociality might be hard to lose. An advanced hymenopteran colony is a highly efficient, well-tuned operation, with elaborate divisions of labor and individual self-lessness far surpassing those in even the most social of human communes. Thus, any reversion to solitary reproduction simply may be infeasible for the individual hymenopteran, regardless of how tempting this personal strategy otherwise might be. In short, rugged individualism no longer may be possible within any eusocial hymenopteran species that long ago got stung with extreme self-sacrifice during the evolutionary process.

Reptiles Whose Sex Is
Temperature Dependent The Greek philosopher Aristotle was also an avid student of nature. However, he entertained several biological notions that we might find downright silly today, such as the nonchangeability of species, or the description of life's complexity as some sort of natural ladder. Another of Aristotle's notions was that the female sex is inherently cold and males are hot. Thus, when warm winds blow, most conceptions are of males, but cold winds produce female embryos. Pretty crazy, huh? Well, not necessarily.

Believe it or not, some species do have temperature-dependent sex determination. Indeed, this is a common situation in many reptiles. Take marine turtles, for example. These animals spend most of their lives at sea, the only exception being when females lumber ashore to lay their eggs in the bottom of sandy pits they laboriously excavate with their flippers. When the babies hatch about eight weeks later, most will be males if the temperature in the sand was low, and most will be females if the temperature was high. At intermediate temperatures, both sons and daughters arise such that the sex ratio in a clutch is more nearly 1:1. A

similar situation pertains for many other turtle species as well, such as the painted turtle *(Chrysemys picta)* of eastern North America. Each female lives in a pond or stream but once or twice a year comes ashore to dig a pit and lay her eggs in a bank of soil.

Before the temperature-dependent nature of sex determination in turtles was known, well-intentioned conservation biologists sometimes dug up the nests of endangered and other species and incubated the eggs in shady enclosures fenced to exclude predators. Unbeknown to the biologists, this practice produced mostly male turtles for later release, a counterproductive outcome with respect to the goal of enhancing subsequent turtle reproduction. The young turtles couldn't be sexed by external examination, so the mistake went unnoticed for a long time, until someone finally thought to examine the internal gonads of the hatchlings.

For some reptilian species, cooler nest temperatures tend to produce female offspring rather than male. In alligators, for example, all eggs incubated at 30°C or less develop into females, whereas temperatures of 34°C or higher result almost exclusively in males. So, if Aristotle had been thinking of alligators rather than humans (or turtles), his notion that hotter winds produce males might not have been so absurd after all.

What have genetic markers contributed to our understanding of sex determination in reptiles? Absolutely nothing. In this case, there are no genetic markers for gender because the sex of an individual is determined environmentally rather than genetically. Males produce sperm and females produce eggs, but in regard to genetic makeup the two sexes are alike. The topic of temperature-dependent sex determination is included in this book merely because of its interesting departure from the various gene-based sex-determining mechanisms characteristic of most other species. Its inclusion also serves notice that not all questions about nature's ways can be informed by the application of genetic markers.

5

Unusual Mating Practices

This part will illustrate how molecular markers have provided even more intimate peeps into nature's boudoirs by laying bare the sometimes hedonistic performances of several creatures with peculiarly memorable sexual practices. In nearly every case, the primary question of exactly who mated with whom to produce the progeny of interest is answered by the genetic markers. This question usually is tackled by conducting genetic paternity or maternity analyses on offspring, one at a time, thereby establishing each individual's biological parents. From an accumulation of such genetic evidence for many families, the "genetic mating system" of a population is illuminated.

Genetic mating systems are of several types. Monogamy refers to the situation in which individuals breed as pairs, each male and female forming an exclusive mating relationship. Polygamy is when individuals breed with more than one partner, and it comes in two mirror-image forms: polygyny, in which many or all males breed with multiple females, and polyandry, in which many or all females breed with multiple males. When both genders typically breed with several partners, the mating system is termed polygynandry or promiscuity.

An important point is that the genetic mating system of a species can differ from the "social mating system" (behavioral interactions and matings, typically as observed in field studies). This discrepancy usually arises from the fact that copulations and mating behaviors in nature tend to be rather secretive affairs, and not all copulations result in offspring. In various animal populations, appreciable fractions of offspring have proved to result from sneaky mating behaviors that remained hidden from direct view, before cast into the open light by genetic parentage analysis.

For example, genetic analyses have revealed a far higher frequency of "cuckoldry" in many avian species than formerly was supposed. A cuckold is a male whose primary mate has proved unfaithful, such that he may not be the true sire of his purported children. He is said to have been cuckolded by the cuckolder. In many species, the elaborate cuckoldry behaviors by these cheating females and their gigolos have been raised to remarkable heights of artistic (if deceitful) endeavor.

Some riveting instances of cuckoldry in the animal world will be examined in this part, as well as several other intriguing questions concerning who has mated with whom, how often, and what the social and biological consequences can be.

Fatherly Devotion and
Female Impersonators In most fish species, females lay their eggs into the outside watery environment. Males then release sperm over the eggs such that fertilization is external (rather than internal, as in

female-pregnant mosquitofish ["The Onus of Pregnancy" in Part 4] or male-pregnant seahorses ["Male Pregnancy," also in Part 4]). Nonetheless, even in externally fertilizing species, parents can be highly devoted to their children. Indeed, more than 20 percent of the eighty-four taxonomic families of bony fishes include at least some species in which adults provide extended care to the offspring, and in about 80 percent of those cases it is the male who devotes the majority of effort to rearing a clutch. This contrasts sharply with the situation in most other vertebrate groups with parental care, where the mothers normally are more doting.

Paternal care in fishes sometimes takes the form of oral brooding, wherein a father gathers and holds the offspring in his mouth for protection. Many bullheads and catfish display this behavior. Another form of paternal care involves building and tending nests. Consider, for example, the dozen or so species of North American sunfish in the genus *Lepomis*. Each spring, a breeding male scoops out a depression along the shallow margin of a pond or creek. After a female lays dozens or hundreds of eggs into his nest, the male squirts on sperm and tends the eggs, each of which soon contains an embryo. For the next couple of weeks, the devoted father hovers nearby, chasing away predators and intruders and also fanning the nest to keep the developing babies aerated and free of silt. The eggs then hatch into free-swimming fry, who leave the relative security of their home.

External fertilization coupled with male parental care opens an ecological window of opportunity for other males to steal some of the fertilization events and thereby surreptitiously commandeer a portion of the reproductive effort by the resident nest-tender. In other words, the possibility of cuckoldry arises. But prior to the use of genetic markers, there was no way to be sure if this additional route to paternity actually was utilized. Are nest-tending sunfish males indeed cuckolded, and if so, how frequently and by whom?

Based on genetic parentage analyses of one well-studied population of redbreast sunfish *(Lepomis auritus)*, two to six females on average spawned in a male's nest, and nearly 5 percent of the resulting progeny were not sired by the nest-tending male. These latter offspring resulted from sneaked or stolen fertilizations by nonresident males who apparently had released sperm onto the nest during spawning bouts by the guardian male and his mates.

Another sunfish species, the bluegill *(Lepomis macrochirus)*, has taken the art of cuckoldry to new depths. This handsome purple-toned species is a popular panfish, of special interest to scientists because of its extensive reproductive repertoire. Resident or "bourgeois" bluegill males, usually seven to eleven years old, build their nests in dense colonies. Each brightly colored male patrols the boundary of his own nest, fiercely defending against unwanted intruders, which include nearly everything except potential mates. The females, whom he hopes to lure, are smaller and duller in color.

Added to this melee is another kind of reproductive player, the cuckolder, of which at least two types exist. Precocious cuckolder males, just two to three years old, dart into a nest while the bourgeois male is preoccupied by spawning with an egg-releasing mate. Like sex-crazed teenagers, these impertinent "sneakers" release sperm too, and then quickly retreat from the nest. Older cuckolder males, called satellites, are female impersonators, retaining throughout adult life the behavior and coloration typical of females. These "female mimics" thereby deceive a bourgeois male and gain temporary access to his nest, only to participate in a ménage à trois (or more) by releasing sperm when bona fide females arrive to spawn. In a few cases, bourgeois sunfish males themselves may be cuckolders, coveting their neighbors' mates and occasionally stealing a few fertilizations while briefly visiting nearby nests.

By revealing actual paternity within clutches, genetic markers have helped to sort out the consequences of this reproductive circus. In several bluegill colonies genetically surveyed in Ontario, Canada, the frequencies of offspring resulting from cuckoldry ranged as high as 60 percent. Furthermore, cuckoldry rates proved to be correlated with the local densities of potential cuckolders (mostly the young males and the female impersonators). Thus, in sunfish, cuckoldry as a reproductive strategy indeed can be a successful alternative to honest parenting. For bluegill sunfish males in particular, both sneaky behaviors and cross-dressing can have their genetic rewards.

Female Accomplices
of Male Cuckoldry
In most fish species where cuckoldry occurs, it is primarily a male-to-male affair— sneaker or satellite males rather directly usurp a portion of the reproductive effort of nest-tending males by surreptitiously fertilizing some newly laid eggs in the guardian male's nest (see previous essay). Cuckoldry is known to occur in an avian species too, the spotted sandpiper (*Actitis macularia*), but in this case there is an added twist: legitimate females (real females, rather than female-impersonating males) mediate the illegitimate process. They do so via "sperm storage."

In many bird species, females are known to store sperm from earlier matings. For example, domestic chicken and turkey hens often produce fertile eggs four to ten weeks after copulating with a male, indicating that these females have capabilities for successful long-term storage of a male's viable sperm. Indeed, in some avian species the female reproductive tracts have special sperm storage glands (as do those of many other vertebrate and invertebrate species with internal fertilization [see "The Storage of Sperm by Females," in Part 6]). In the spotted sandpiper, such sperm storage has some unexpected reproductive consequences for late-mating males.

Spotted sandpipers are a familiar sight across Canada and the northern United States during the summer months, when they bob and jig like little hip-hop dancers while patrolling freshwater shorelines. Like several other shorebirds, this polyandrous species (in which some females have multiple mates) tends toward sex-role reversal. Following spring migration northward, aggressive females arrive on the breeding grounds about a week before the males, and fight to establish territories to which males are attracted. Over several weeks, each female copulates successively with as many as four different mates and then lays clutches of eggs that the respective males tend. The males in this species provide nearly all the parental duties of egg incubation and brood care.

A remarkable finding from genetic paternity analyses of spotted sandpipers is that the clutches tended by late-mating males sometimes contain babies conceived by sperm from an earlier mating. Thus, a male who pairs with a female early in the mating season cuckolds his female's later mates by means of sperm that he has deposited in her body. In other words, a female may use stored sperm, rather than fresh sperm from her latest mate, to fertilize some of her eggs.

For the cuckolded male, this is a reproductive double whammy. Not only is he burdened with heavy parenting duties, but his time and energy are invested in rearing the biological progeny of a rival male. This hard-working but unwitting foster father has been duped reproductively by his mating partner. In contrast, the cuckolder doubly benefits. Not only does an early mating male have greater confidence of genetic paternity for his own brood, but he also enhances his own personal genetic fitness by appropriating the parental efforts of his unfaithful mate's subsequent partners.

Lizards That Play
Rock-Paper-Scissors

The side-blotched lizard *(Uta stansburiana)* is a trim, handsome little creature abundant in sandy regions, desert flats, and arid foothills of western North America. The species gets its name from an inky blue-black spot just behind the armpit in both sexes. Otherwise, the females are tan colored, delicately striped, and spotted. The males, however, come in three alternative types that differ not only in appearance but also in mating strategy.

One form of male has a blue throat, is territorial, and guards its mate. Another form with an orange throat is hyperterritorial and polygynous, avidly mating with multiple females. A third form is yellow-throated and does not regularly hold or defend turf. Instead, it gains access to the territories of defender males by mimicking females and, once present, "sneaking" copulations with some of the resident females. Field observations led to the hypothesis that the joint persistence of these three distinctive types of males may be explained by analogy to the popular children's game of "rock-paper-scissors," in which a rock can smash (trump) scissors, scissors can cut paper, and paper can cover rocks.

It seems that via their proclivity to guard mates, blue-throated males (think of them as "paper") usually avoid cuckoldry by yellow-throated sneaker males ("rocks"), and thereby effectively cover, or trump, the rock strategy. However, this mate-guarding tactic is ineffective against the more aggressive orange-throated males ("scissors"), who can cut a blue-throated male's paper strategy to shreds by mating with his females. Yet, territories of the hyperaggressive orange-throated males sometimes get

too large to manage, so they are particularly susceptible to cuckoldry by yellow-throated sneaker males (rocks smash scissors). Thus, this triad of behavioral interactions with regard to mate acquisition means that no one mating strategy by males would seem to have the clear upper hand.

Do the actual patterns of biological paternity, as documented by genetic markers, conform to these predictions and thereby support the rock-paper-scissors scenario for side-blotched lizards? In short, yes. For example, a large proportion of the cuckoldry against blue-throated males did indeed prove to be by orange-throated males, and a high fraction of the cuckoldry against orange-throated males was by yellow-throated males.

Another interesting phenomenon uncovered in the genetic analysis was posthumous fatherhood. This occurred with regularity because a female side-blotched lizard apparently can store and utilize the sperm from an earlier mating to fertilize some of her eggs, even long after her children's true biological sire may have passed away. Perhaps this means that a fourth morpho-type of male could be recognized. Such males themselves may be long gone, but their reproductive potency lives on, stored as viable sperm in the inner recesses of their surviving mates.

Birds with Roving Eyes and Cheating Hearts

Most "perching birds" (Passeriformes) are either little brown birds or brightly colored ones that flit about in trees and bushes. Traditionally, they were thought to be among the most monogamous of organisms. Males and females of most of these species pair off during each nesting season in what seemed to be faithful couples fully devoted to each other and to the tasks of building and defending a home and raising chicks. This conventional wisdom has turned out to be a Pollyanna view.

The truth is more interesting. Many and perhaps most species of perching birds routinely engage in "extra-pair" (extra-marital) affairs, such that a substantial fraction of nests contain at least some progeny not sired by the male who tends the clutch or guards the territory. This knowledge has come to light from genetic paternity assessments, which have revealed that nestlings do not always carry the male's genes.

One case in point involves the eastern bluebird *(Sialia sialis)*, a familiar sight in North American pasturelands. In some populations, according to genetic evidence, more than one-third of the chicks in nest boxes were not sired by the territorial male who by all other appearances was the father. Thus, there must be a lot of extra-pair mating activity that had gone unnoticed by field observers. Subsequent analyses revealed that younger males are cuckolded more frequently than older males, as are males whose female partners commonly strayed from the defended territories during their fertile periods. It now appears that female bluebirds actively pursue copulations with foreign males and that these males often oblige.

Another perching bird with similar genetic surprises is the dunnock *(Prunella modularis)*, a small unobtrusive sparrow of European parks and hedgerows. Drab brown in appearance and of a retiring nature, dunnocks are socially polyandrous, meaning in this case that a female often consorts with two males during the breeding season. An alpha male is her primary escort, but in the bushes nearby lurks an ever-present beta male, the female's paramour on the side. Both males defend the trio's territory, but the alpha male is dominant and continually shoos away the beta male. However, this doesn't stop the female from sneaking off at the least opportunity to mate on the sly with the beta male in the hedges.

Dunnock copulation (as in most birds) is a quick affair, taking less than a second. Sperm is transferred from male to female in a "cloacal kiss" as the genital openings of the partners come together for an instant. Dunnocks copulate frequently during the mating period, about twice per hour. The most intriguing aspect of dunnock mating, however, is a prolonged pre-copulatory behavior by the alpha male. When his mate returns from a liaison with the beta male, the alpha male hops behind her and, as she quivers her wings and raises her tail, he pecks at her exposed cloaca about thirty times. In response, the female's cloaca becomes pink and distended, and then with strong pumping movements ejects a small drop of fluid. This droplet is full of sperm, presumably from beta. Alpha and his mate then copulate.

Who, then, fathers the chicks in these dunnock trios? Genetic markers have revealed that the alpha and beta males each sire on average about one-half of the nestlings produced by their female. Thus, the pre-copulatory love pecks by the alpha male that cause his mate to eject stored sperm are less than fully effective in thwarting parenthood by beta.

Treefrog Mating Ceremonies
Treefrogs (genus *Hyla*) are amphibians with several peculiar features: adhesive disks on the tips of widely splayed toes, enabling them to cling to branches and even walls and windows; a vocal sac under the chin that a male inflates into a round balloon during courtship (as if blowing a big bubble with chewing gum); and an ability to alter skin color somewhat to match the environment. About ten treefrog species inhabit the wetlands of eastern North America.

Two common forms are the green treefrog *(H. cinerea)* and the barking treefrog *(H. gratiosa)*. The former has a trim body with smooth, brilliant green skin, yellowish racing stripes down both sides, and golden spots on its back. The male's voice is a cowbell-like "queenk-queenk-queenk" with a nasal inflection, repeated dozens of times per minute on muggy nights throughout the spring and summer breeding months. The barking treefrog, by comparison, is fatter, with a rough-hewn, gray-green skin profusely spotted with dark round blotches. This species gets its name from the male's raucous barks, often issued from the treetops. The breeding call, however, typically is issued from the surface of a pond. It is an explosive "doonk" or "toonk," repeated at one- or-two-second intervals.

Normally, these two species are readily distinguishable. However, at a few sites, such as in some farm ponds near Auburn, Alabama, individuals with a stout body form like the barking treefrogs but a skin pattern more reminiscent of green treefrogs are observed as well. Apart from being intermediate in appearance, these frogs also have strange "guoonk-guoonk" voices, and, thus, they were somewhat of a scientific mystery. Were these curious specimens hybrids, and if so, how had the matings taken place?

Genetic markers in conjunction with field observations have provided the answers. The odd-looking, weird-sounding individuals at the Auburn site proved to be hybrids between barking treefrogs and green treefrogs. Thus, each hybrid specimen contained mixtures of various genetic markers from the two parental species. A more startling genetic discovery (based on maternally inherited mtDNA) was that nearly all the hybrids had barking treefrog mothers (and, hence, green treefrog fathers) rather than the reverse set of parents. Why might this be the case? It probably has a lot to do with the stereotypic mating behaviors of the two species.

In treefrogs, mating occurs when a male jumps on a female's back, digs his front toes into her sides, and hangs on for dear life until she lays her eggs in the water, which he then fertilizes. In general, treefrog males are avid maters, eager to mount nearly anything their size that hops. But there is an interesting prelude to this activity that undoubtedly accounts for the pronounced sexual asymmetry in how the hybrid frogs were produced at the Auburn site. Typically, barking treefrog males issue their mating calls while floating on the surface of a pond; green treefrog males sing from grasses or

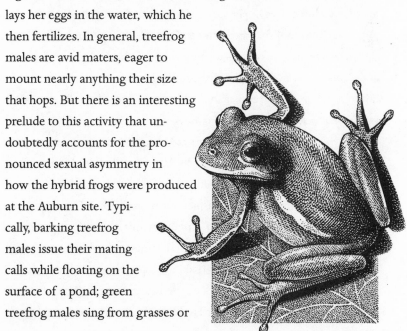

low brush along the shoreline. During the day, females of both species lounge in adjoining woods, but hop down to the ponds at night in response to the alluring male choruses. This physical arrangement of participants in the frogs' mating-pond chapel is the key to the asymmetry in the hybridization process.

Because of the distinctive calling perches of the males, female green treefrogs seldom encounter barking treefrog males on their nuptial travels from the woods to the pond. However, female barking treefrogs must "hop a gauntlet" of green treefrog males encircling the shoreline before reaching the pond-floating suitors of their own species. Thus, behavioral evidence in conjunction with the genetic data suggest that green treefrog males lurking in the grass occasionally intercept and mate with the barking treefrog females. By contrast, the physical arrangement itself means that barking treefrog males on the pond surface seldom have the opportunity to mate with green treefrog females.

Swordtails' Tales

Swordtails and platys (genus *Xiphophorus*) are familiar creatures in pet stores. More than twenty species of these popular aquarium fishes are native to Central America, and collectively they display a palette of eye-catching colors. In body shape and overall appearance, however, all these fish look quite alike, the most noticeable difference being the presence in adult male swordtails of an elongated tail fin that resembles a pointed cutlass.

The male swordtail's sword is particularly interesting to scientists and, apparently, to the female fish as well. Why would males have evolved such a burdensome apparatus that seemingly would impede swimming and make the animals more conspicuous to predators? The answer resides in the mating behaviors of female swordtails and platys. As has been shown in several experiments, these females clearly prefer to mate with saber-adorned males. Remarkably, this preference holds true even for platy females, whose males lack the rapier-like tail. Thus, when a male of their own species is given an artificial sword (by surgically grafting a plastic prosthesis to his tail), female platys consistently prefer this male to others.

These experimental findings helped give rise to the "preexisting bias" hypothesis, which posits that female *Xiphophorus* fish had an innate mating preference for adorned males that predated the evolutionary elaboration of the sworded condition. One key prediction of this hypothesis is that swordlessness is the ancestral (or primitive) state of affairs for this group of fishes, such that these preexisting mating preferences of females for sworded males has driven, where possible, the evolution of male swords.

This intriguing hypothesis has been tested genetically. Using numerous molecular markers, scientists recently deciphered the evolutionary (phylogenetic) relationships for twenty-two species of platys and swordtails. By then plotting the presence versus the absence of male swords on this gene-based phylogeny, they deduced that the common ancestor of this entire taxonomic group must itself have possessed a sword. This surprising finding initially called into question the "preexisting bias" hypothesis for the evolution of sworded tails, because it indicated that the sworded condition had arisen much earlier than previously thought.

However, another surprise was soon in store. Further experiments demonstrated that females of some species outside the genus *Xiphophorus* also tend to prefer males with artificial tail adornments. The ances-

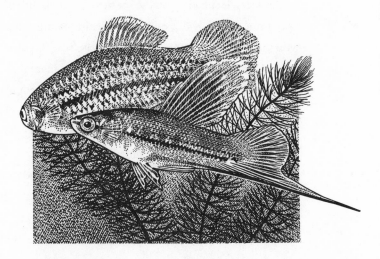

tral condition for these other fish almost certainly was swordlessness. Thus, new life was breathed into the preexisting bias theory, which remains alive, at least for now.

The genetic findings also raised another puzzling question. If the presence of swords was the ancestral condition for *Xiphophorus,* and if female platys and swordfish universally prefer males with swords, why was the cutlass tail lost in various platy lineages? The answer likely involves the notion that swords, although attractive to females, are cumbersome for males to lug around and probably decrease their mean survival (for example, by increasing the costs in energy or risks of predation).

Thus, in the evolutionary lineages leading to modern-day platys (but not swordtails), males without swords must have been favored, on balance, by natural selection. Apparently, in nature's evolutionary wars, swords confer both fitness benefits and costs to their bearers.

Why Some Species Like Leks

Within a small but taxonomically diverse array of creatures, ranging from a few mammals to miscellaneous species of amphibians, fishes, and insects, groups of males aggregate at traditional sites where they fervently display in hopes of attracting receptive females. These "lekking sites" are nature's version of singles' bars. Depending on the species, males on the lek may sing their hearts out, dance avidly, release mate-stimulating chemicals, or otherwise vie for the attentions of the opposite sex. Females frequent these singles' bars to peruse the goods and choose a mate. Usually, only a few of the most resplendent males achieve most of the copulations. So, why do subordinate males participate in this mating ritual? Three traditional ideas (not mutually exclusive) all suppose that lesser males actually do increase their mating opportunities by joining a lek.

Under the "hotspot" hypothesis, males simply tend to aggregate at female hangouts, for example, near rich feeding grounds. Thus, the groups are merely a by-product or consequence of the distribution of environmental resources. Under a second hypothesis, the "hotshot" model, males purposefully display near other males to increase their chances of being noticed by the females. Under a third hypothesis, known as the "female-preference" model, males aggregate because females prefer to visit a crowded lekking site for purposes of quick and safe mating. Under any of these models, even subordinate males occasionally may get lucky. Thus, each of these notions focuses on possible mating benefits to males who join leks.

However, two recent genetic studies have provided unexpected evidence for yet another selective mechanism that might help to explain the evolution of lekking behavior in at least some species. Many chickenlike birds (family Phasianidae) have spectacular leks. For example, in both the European black grouse *(Tetrao tetrix)* and the peafowl *(Pavo cristatus)*, males on the lek display their ostentatious feathery finery in great strutting rituals. The females profess disinterest, but they soon mate, often with the most splendiferous males.

For each of these two species, the unexpected genetic finding is that males within a lek are close kin, with an average genetic relatedness like that of half-brothers. Indeed, many of the subordinate males on the lek appear to be brothers or sons of the alpha male. This means that even if

an alpha male conceives most of the children, these may be nieces, nephews, or even siblings of other males on the lek. Thus, by helping to attract females to their own rather than another lek, subsidiary males can enhance the genetic fitness of the group via kin selection. In other words, a subordinate male will see copies of his genes transmitted to the next generation through relatives, even if he himself seldom wins a mate.

How it is that closely related males come to aggregate on these leks is currently under investigation. In the black grouse, it simply may be a matter of the faithfulness of males to their natal sites. In the peafowl, more active behavioral mechanisms of kin recognition seem to be at play.

The Naked Mole Rat
What creature is nearly blind, naked, looks like a fat sausage with wrinkly skin, lives in large subterranean colonies, and makes a habit of eating the feces of its fellows? No, it's not some abominable beast conjured up for a science fiction story. It's the real-life naked mole rat, *Heterocephalus glaber*, of central Africa. Through a naturalist's eyes, this outwardly homely and disgusting creature is actually quite beautiful for the scientific schooling it provides.

Naked mole rats live in underground tunnels, in close-knit groups of 30 to 300 individuals. Within each colony, only one female (the queen) and her one to three mates reproduce. All other colony members are nonreproductive workers or helpers, typically the young from previous

litters. They assist the queen and her mates in rearing the colony's babies and in maintaining and defending the burrow system.

Thus, in many respects, naked mole rats are a mammalian analogue of ants, bees, and wasps that likewise live in "eusocial" communes. In each case, worker individuals forfeit their own reproduction to toil instead in behalf of the colony and its royalty. As described earlier ("Extreme Social Behavior and Gender Control," in Part 4), the evolution of eusociality in social insects is related to their unique haplo-diploid basis of sex determination. So, how did naked mole rats (diploid creatures with standard genetic means of producing males and females) also come to display such extreme degrees of reproductive selflessness and worker subordination?

Colonies of the naked mole rat recently have been examined using DNA fingerprinting techniques. The most important finding is that colony mates proved to be strikingly similar genetically, by mammalian standards. This suggests that inbreeding within a colony is intense in this species. Indeed, genetic calculations indicate that most of the babies within a mole rat colony were conceived of matings between siblings or between parents and their older offspring.

Normally, in other mammal species, close inbreeding is harmful, leading to the production of genetically deformed or disabled progeny. Nonetheless, inbreeding may have been the best evolutionary option available to naked mole rats. In their predator-rich African homeland, these waddling sausages would be tasty morsels if they left their home burrows. The patchiness of the animals' food (large tubers in addition to feces), and the difficulty of digging burrows in the baked African soils, must further have favored stay-at-home behavior. However, by remaining in the colony in which it was born, a mole rat's reproductive options become limited. Inbreeding ensues, and with it a stronger opportunity for nepotism (favoritism toward kin). Perhaps only some fraction of the original colonies survived this initial phase, but those that did eventually became purged of inbreeding's harmful effects and began to flourish.

As in the social insects, over the eons this must have led to the evolution of extreme helping behavior that is the hallmark of eusocial reproduction. Thus, the evolution of cooperative breeding and eusociality in this homely little mammal is related to an interplay between ecological and genetic factors. Although much remains to be learned about the biology of this remarkable species, naturalists and geneticists have begun to strip at least some of the cloak of mystery from the naked mole rat.

6

Novel Ways of Handling Eggs and Sperm

The reproductive repertoires of many species include clandestine or otherwise fraudulent mating practices and sexual behaviors by certain males or females. This part will highlight more examples of this sort, in this case involving some of the unusual ways in which organisms manipulate sperm, pollen, or eggs either before, during, or following a mating event. In various species these maneuvers include multiyear storage and use of sperm by females, postmating choice of sperm or pollen by females, sperm removal following copulation, pollen-tube races, furtive egg-dumping, egg thievery, nest piracy, egg consumption by parents, and even the display of egg-mimic features on the body parts of males. In some cases, genetic markers have documented these behaviors for the first time, and in all cases they have greatly illuminated how, when, and where these phenomena operate in the wild.

Egg-Dumping and Wily Cuckoos

Not all chicks in a bird's nest are necessarily the biological offspring of the adults who feed and tend them. How can this be? Often, this happens when a female lays one or a few eggs in someone else's nest, leaving the subsequent parenting duties to an unsuspecting foster couple. When both the perpetrator and her gullible targets belong to the same species, this reproductive sting operation is known as "intraspecific brood parasitism," or "egg-dumping" for short. Usually, the stealthy egg dumper sneaks into a nest while the owners are out, quickly lays an egg, sometimes removes one of the couple's own eggs (so they may not notice the deed when they return), and then quickly retreats.

One species in which egg-dumping has been documented genetically is the house wren (*Troglodytes aedon*). This tiny brown denizen of New World gardens is known best for its beautiful, elaborate song, and for a perky, up-tilted tail that seems to reflect the bird's impertinent attitude to life. Its generic name comes from "troglodyte" (cave dweller), and refers to the species' habit of nesting in tree cavities or nest boxes. In one population studied genetically, at least 2 percent of the chicks were documented to have biological mothers other than the female who was rearing them.

Before the advent of molecular genetic techniques, egg-dumping in several avian species had been suspected from direct behavioral evidence. Observers noticed, for example, that eggs sometimes appeared in a nest outside the normal laying sequence of the nest's owners, or that they looked a bit different from other eggs in the clutch. Genetic markers merely confirmed suspicions that such eggs from a reproductive parasite indeed can result in foster children.

The common cuckoo (*Cuculus canorus*) of European gardens and forests has taken the egg-dumping strategy to new heights. It has abandoned dutiful parenthood altogether and instead lays its eggs exclusively

in the nests of other bird species, a phenomenon known as "interspecific brood para-
sitism." The duped foster parents are left with the job of incubating a
cuckoo's egg and rearing the chick. Because cuckoos are rather large
birds (about the size of a small hawk), yet are reproductive parasites on
much smaller species, rather ludicrous feeding situations are a common
sight. A rotund cuckoo baby may beg and receive food from a foster parent that is only one-half its size.

Common cuckoos have special physical and behavioral adaptations
for brood parasitism. The female has a protrusible cloaca that she ex-
tends into the nest of a smaller species to drop an egg in just a few sec-
onds. While briefly at the nest on her surreptitious mission, she usually
removes and eats one of the host's eggs. The cuckoo egg itself is thick
shelled, resistant to destruction by the host species. After hatching, the
cuckoo chick displays an ejection instinct in which it arches its back and
pushes hard against any object it encounters. Usually, this is a host's egg
or hatchling, which literally gets shoved over the rim of the nest to fall
and perish. The baby cuckoo has a huge mouth with a red gape that elic-
its an automatic feeding response by the foster parents. The cuckoo chick
also exhibits continuous and exaggerated food-begging calls.

The common cuckoo has yet another trick up its sleeve: "egg mimicry." Cuckoos tend to specialize on different host species, for example

on redstarts and winchats (which lay blue eggs) in Scandinavia, and on reed warblers (which lay gray-green eggs blotched with black) in central Europe. Usually, the color of the egg that a cuckoo lays into a nest closely matches that of its host. Each cuckoo population that lays eggs of a particular color pattern is referred to as a "gens" (plural, "gentes").

Do cuckoo gentes represent divergent genetic "races" or, perhaps, even different species? In the populations surveyed for mtDNA and other gene markers, each cuckoo gens appears to be a distinct female lineage, albeit closely similar in overall genetic makeup to other such gentes. These genetic findings imply that female lineages of European cuckoos specialize on particular host species, yet have remained part of the same species by mating freely with common cuckoo males of different types. The genetic results also indicate that egg mimicry can evolve rapidly (in response to strong selection pressures by discriminating hosts) and that shifts in hosts have occurred frequently and recently during cuckoo evolution.

The common cuckoo is among a small handful of avian species around the world that forgo devoted parenthood and rely exclusively on the unpaid incubation and child-rearing services of other birds. Are the victimized foster parents entirely helpless against brood parasitism? Not necessarily. Among the counter-adaptations displayed by one or another potential host species are the following: nest guarding and mobbing behaviors that inhibit egg-dumping by a sneaky brood parasite; desertion or reconstruction of a nest in which a parasitic egg has been deposited; destruction of the parasitic egg (hence the selection pressure for egg mimicry by a gens); and refusal to feed a foster child. Such countermeasures are less than fully effective, however, because cuckoos and other brood parasites continue to succeed.

For the selfish brood parasite, egg-dumping would seem to be a great strategy if she can get away with it. From the foster parents' genetic perspective, it's a bum deal. To the foster child, it probably doesn't much matter who the attentive adults are. And to a human observer, it's just another example of nature's endless ingenuity in getting genes into the next generation.

The Nest Architecture
of Swallows

The bird family Hirundinidae is made up of about ninety species—the swallows—that are among nature's most accomplished and graceful flyers. These gregarious, day-active birds spend most of their time on the wing, plucking insects from the skies as they fly over meadows and lakes in lovely aerial ballets. In the spring and fall, most of these birds also migrate great distances, feeding on the fly, in cross-continental journeys that they seem to manage as effortlessly as humans stroll in a park. For example, trim barn swallows grace North American farmlands in the summer, then fly south in the autumn through Central America to their winter quarters in South America.

So aerial are the swallows that their feet have become small and weak, and they walk only with difficulty. Yet return to the ground they must, to build and tend their nests each summer. The varied architectures of their nests also make these birds particularly fascinating. Among the dozens of taxonomic families of songbirds, swallows display the greatest diversity of nest construction modes.

Some species, such as the bank swallow, burrow into vertical cliffs, depositing their eggs and rearing their young in the safety of a cuplike cavity at the end of a narrow tunnel through the soil. Other species, such as the tree swallow

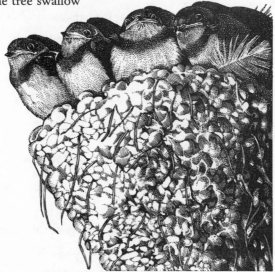

and purple martin, use adopted natural cavities in trees (or, nowadays, nest boxes provided by humans). Still other species construct nests of mud, and here too lies great variety. The barn swallow makes and attaches an open-ceiling mud cup to a vertical surface; the house martin puts a roof of clay over its nest, which resembles a dried-mud goblet; and the cliff swallow adds an overhanging eave to its adobe nest.

Before the modern era of genetic assays, a general supposition was that nest construction behavior in swallows is evolutionarily "plastic," easily and readily moldable as the birds adapt to local circumstances. Thus, it came as a great surprise to learn that the varied architectures of swallows' nests matched very closely the species' phylogenetic relationships as gauged by molecular genetic data. In other words, it turned out that all cavity-adopting species, such as the tree swallow, are on one branch of the extended swallow family tree, and the same is true for each of the various types of mud-cup nesters.

By mapping the various modes of nest building onto the phylogenetic tree, it was also possible for scientists to deduce the evolutionary order of appearance of these nest-building behaviors in swallows, and to decipher how nest architecture in this taxonomic group had evolved over the eons. From such phylogenetic analyses, it appears that burrowing into the soil is the primitive or ancestral condition (still retained by some living species, such as the bank swallow). From this initial condition, cavity nesting then evolved in one major branch of the swallow family tree, and mud-construction methods arose and then diversified in another.

Egg Thievery and Nest Piracy

Stickleback fishes (family Gasterosteidae) get their name from a series of hard dorsal spines that, together with sharp spurs projecting from some other fins, greatly discourage predators. A row of bony plates in lieu of scales along each side of the torso completes the stickleback's defensive body suit and gives these bristly fish the formidable appearance of miniature knights in armor. Yet each fish is only an inch or two long. About eight living species inhabit fresh- and saltwater environments in high northern latitudes.

Most male sticklebacks build and tend elaborate nests. Each male constructs his tunnel-like home from loosely woven aquatic plants glued together with special secretions from his kidney. He then attracts a passing female and lures her into his nest by a herky-jerky dance. He fertilizes the eggs that the female lays and then chases her off. Occasionally, one or more sneaker males appear on the courtship scene and may dart through the nest tunnel too, releasing sperm. The resident male may spawn in close succession with two or more females, thereafter tending the resulting embryos until they are ready to leave home as free-swimming fry.

Genetic parentage studies of the fifteen-spined stickleback *(Spinachia spinachia)* in Sweden revealed that a male's nest frequently contained embryos from multiple females, and that many of the progeny were fathered by sneaker males. Of greater surprise was the genetic documentation of egg thievery. Biologists long had suspected that males occasionally pilfer eggs from other nests, because they can be seen toting clumps of eggs, in their mouths, back to their own homes. By firmly documenting the presence of foster offspring in various nests, the genetic markers helped to confirm that nest-tending males sometimes steal a batch of fertilized eggs from another male.

Why would a male steal eggs from a rival, thereby voluntarily assuming the role of a partial foster father? At face value, it makes no evolutionary sense. Yet, the same puzzling behavior had been noted in other stickleback species, such as the three-spined stickleback. Several speculations have been advanced. Perhaps a male somehow can distinguish his own biological offspring from

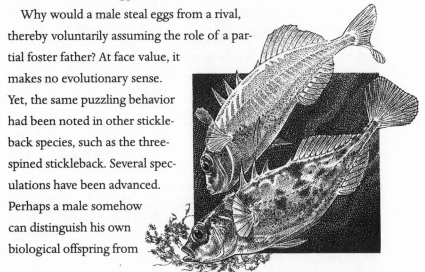

others and stocks his larder with foster children that he gradually will cannibalize over several days as handy snacks (see the last essay in this part). Another possibility is that foster children in the nest provide a pre-dation-dilution effect that on balance benefits the male's own offspring. Under one version of this hypothesis, a predator who discovers and attacks the stickleback's nest might become satiated before eating all of the male's own progeny.

Another behavioral hypothesis is perhaps more plausible. In several fish species, it is known that a female prefers to spawn in nests that already contain eggs. This may be because such eggs are an indicator of good health in her potential mate, or they may speak favorably of his parenting skills. Thus, for stickleback males, occasional egg thievery and foster parentage might make evolutionary sense after all. By stealing a few fertilized eggs, a male may be priming a reproductive pump that brings additional unfertilized eggs into the nest from females duly impressed with his parenting potential.

Foster parentage by nest-tending males is called "allopaternal care." Apart from egg thievery, another pilferous route to allopaternal care in some fishes is "nest piracy," wherein a male sometimes adopts a nest built by another male of the same species. This may be a non-confrontational "nest takeover" event after the former resident has died or deserted, or it may involve an active eviction of the former resident by an aggressive intruder. In any event, nest piracy has been documented genetically (by molecular parentage analyses) in several nest-tending species, including sunfish, sand gobies, and darters. It may be especially common when favorable nesting sites for males are at a premium.

To a thief who steals food or other such commodities from another fish, larceny has a certain logic. By contrast, stealing fertilized eggs (or pilfering nests that contain fertilized eggs) seems idiotic for the thief, at face value, because the filcher then ends up rearing a competitor's offspring. Nonetheless, if egg thievery and nest piracy also win for pirates the favor of gravid (egg-carrying) females, then this kind of larceny too has its reproductive rewards.

Males Whose Body Parts Mimic Eggs

As mentioned in the last essay, females in several fish species have a known preference for spawning in nests that already contain eggs. This behavioral proclivity has been documented in sticklebacks, gobies, darters, minnows, sunfish, and several other taxonomic groups of fishes in which males tend nests. Such choosiness by a female may reflect her inclination to spawn with a male who would seem to have good parenting abilities, one who might do a top job of raising her children. This might be the case if such a male was more attentive at the nest, reducing the risk of predation on her offspring, or if he provided them with better genes. Regardless of the exact reason, the fact that females prefer to spawn in egg-containing nests means that any males who tend such nests are likely to have a reproductive advantage, on average, over those who do not.

Thus, this special kind of mating preference by females will tend to favor any nest-tending male who can display healthy eggs to passing females. Apart from personally attracting his first, critical, mate, how can a male "prime the pump" of female spawning in his nest? Two possible behavioral routes were described in the last essay: egg thievery and nest piracy. In both cases, a male puts eggs on display from a female with whom he himself had not mated. In principle, at least one other potential evolutionary route is available to males: trick the females by displaying fake eggs. Remarkably, this seems to be exactly what happens in several darter species in the genus *Etheostoma*.

More than 100 species of darters live mainly in mountain streams of eastern North America, and a handful of these display the phenomenon of "egg mimicry." These are situations in which males have evolved particular body parts that closely resemble eggs of the species. A male in breeding condition typically excavates a nest cavity to which he tries to attract egg-laying females. While peering out from his nest cavity under a rock or ledge, he actively displays his "egg mimics" to any females in the vicinity.

Depending on the species, these egg mimics occur on various parts of a male's body. In the fantail darters, each spine on the foremost dorsal fin carries one round glob that looks for all the world like a fish egg. In the spottail darters, these globs are on the spiny tips of a dorsal fin situated further back on the male's body, and in another group of darters they are on the pelvic fins near the male's throat. Thus, egg-mimicry structures have evolved at least three times, independently, in various darter lineages. Primitive versions of these fleshy protuberances in ancestral darters may have served to blunt the sharp fin spines that otherwise could puncture some of the eggs that line a male's nesting chamber. Probably only later in evolution did the structures become refined and useful also as egg mimics.

In each of the aforementioned darter groups, the number of egg mimics per male is pretty much constant, because each fish has only so many relevant fin rays. However, in one population of yet another species, the striped darter (Etheostoma virgatum), the number of egg mimics is highly variable (from half a dozen to more than twenty). This is because, unlike the physical egg-mimic structures in the other darter species, these egg mimics are spots of white set off against the darker background of the male's pectoral fins.

Do males with more egg-mimic spots on their pectoral fins spawn with more females, on average, than do males with fewer spots? This might be a logical prediction of the hypothesis that these spots truly serve as egg-mimic devices that successfully attract females. This prediction was put to a critical genetic test based on maternity assignments for the dozens of embryos in the nest of each of several striped darter males with varying numbers of egg mimics. Results generally were consistent with the notion that females prefer to spawn with males displaying more egg mimics.

Perhaps this is an honest display—male darters with many egg mimics indeed may be of better health or higher genetic quality, on average, than others, and merely are advertising that truthful fact to females. However, the egg mimics might instead be a somewhat dishonest display by males, signifying little or nothing about their true parenting abilities.

The Storage of Sperm by Females
The reproductive tracts of many female animals include special internal organs that can store viable sperm for considerable periods of time. For example, for as long as two weeks following a mating, a female fruit fly can use the sperm she received from a male to fertilize her eggs. The same is true for many birds. In mammals, females typically can store and use viable sperm for only a few days at most, but some female bats are able to do so for as long as thirty weeks following a copulation.

Some female lizards can use stored sperm to fertilize their eggs for two months or more. Likewise, several months may intervene between copulation and egg laying in various newts and salamanders. In some fish species that have internal pregnancies and bear live young, females can stockpile viable sperm for three to ten months after a mating. Even more impressive, however, are the endurance champions of sperm storage whose feats of gametic hoarding are measured not in days or months but in years. These include some of the social insects, in which a queen bee or ant continues throughout her life to produce offspring using sperm that she has stored internally since her nuptial flight (see "Mating Champions of the Insect World," in Part II, on page 177).

Among vertebrate animals, the endurance champions of sperm storage are female turtles and snakes, who can save and employ the sperm from a copulation for perhaps five years or more. Such long-term sperm storage first was suspected from the observation that captive females housed in isolation may continue to produce offspring long after they last had been with a male. Unless these progeny arose via parthenogenesis (virgin birth; see "The Lizard That Dispensed with Sex," in Part 2), which seemed unlikely, these females must have employed sperm stored from a carnal liaison that had taken place years earlier.

Although female turtles and snakes appear physiologically capable of long-term storage and use of sperm, do they actually employ this ability in nature? Females isolated in zoos might behave very differently from their free-ranging counterparts, who presumably have regular access to mates from whom they could receive fresh sperm for conceiving their children.

To test the possibility of long-term sperm storage in the wild, genetic markers have been employed. One species studied extensively in this regard is the painted turtle, *Chrysemys picta,* a lovely, smooth-shelled inhabitant of shallow, vegetated waterways in eastern North America. Females mate in the water and later come ashore to lay eggs in nests dug along the banks of a lake, marsh, or stream. The genetic study involved paternity analysis of sequential clutches laid across a four-year span by individual females whose identity was known because physical tags had been applied.

In nearly all cases, each female's successive egg clutches within a nesting season proved to have been fertilized by one batch of sperm. This means that the female must have stored and used her mate's gametes for several months following a copulation. Even more remarkable, for as many as three years a female's successive clutches sometimes had been sired by the same male. Painted turtles are not particularly social creatures, are highly mobile, and number at least 500 animals at the study site where the genetic assays were conducted. These and other aspects of turtle biology make it unlikely that a given female had remated with the same male year after year. Instead, she probably stored and continued to use the male's sperm from that original copulation.

Damsels and Dragons

In many types of animals, females are inseminated by multiple mating partners during a single breeding episode. Thus, sperm from two or more males can be present simultaneously within a female's reproductive tract. Presumably in response to the selection pressures associated with this "sperm competition," males of various species have evolved a great diversity of behavioral strategies that serve to reduce that competition, or otherwise bias the fertilization outcome in ways that increase their probability of being the true biological fathers of most of the resulting kids.

Some commonly observed male behaviors that reduce sperm competition and thereby enhance the copulator's likelihood of paternity include prolonged sexual unions (up to a week in some insects), multiple sexual trysts with the same female, and mate-guarding following copulation. For the focal male, each such behavior diminishes the possibility that sperm from competing males can access the female's eggs. Another male strategy is employed by many species of invertebrate animals, snakes, and mammals: secretion of a "copulatory plug" during mating. This substance blocks a female's reproductive tract from subsequent inseminations.

Another ingenious tactic in nature's mating handbook is displayed by several species of dragonflies and damselflies. During the summer months, these familiar citizens of ponds and streams dance though the air in brisk defense of their territories, on busy excursions in search of food or mates, or sometimes flying together in pairs while united in copulation. It is during these latter events that the males employ a peculiar reproductive tactic that presumably has evolved in response to the challenge of sperm competition. In these species, a male's dual function penis bears recurved horns near its tip. During copulation, at the same time that new sperm is released, these cup-like devices physically remove old sperm from the female's reproductive tract. Most likely, that old sperm, present in special female storage organs called spermathecae, had been placed there when the female had copulated with another male.

Who actually fathers various dragonfly and damselfly broods? From paternity analyses based on genetic markers, researchers showed that a copulator's tactic of sperm removal wasn't always perfect, because these males had not sired many of the babies in their respective mate's broods. It turns out that in many cases a female damselfly or dragonfly can trump the male's sperm-removal tactic by using older sperm from a back-up sperm storage organ, the bursa copulatrix. Unlike the spermathecae, this structure is not scoured out by the pronged penis of the female's current suitor.

Why a female continues to use some of the sperm from a prior mate is not entirely clear from these experiments. However, the findings do suggest that females can play a key role in the sperm competition wars too. Perhaps this really isn't so surprising. After all, each female has a huge vested interest in who sires her children. Indeed, pre-copulatory mate choice by females is nearly ubiquitous in the animal world. Females typically are very choosy, highly discriminating in regard to mating partners. Now, scientists strongly suspect that females in various species have some sophisticated post-mating options as well, by being choosy about which lots of stored and newly acquired sperm will be employed to fertilize their eggs.

Beautiful Iris

Competition among sperm for fertilization success is rampant in the animal world, but plants have their own ubiquitous version too. "Pollen competition" occurs whenever multiple grains of pollen from the stamens (male parts) of different flowers land jointly on the delicate stigmas (female parts) of a flower. If all goes properly, these eager pollen grains quickly germinate, sending pollen tubes down through an elongate feminine tissue (the style) that connects the outer stigma to the inner ovary with its hidden eggs. Thus, all else being equal, pollen grains with faster-growing tubes are more likely to achieve most of the fertilizations. By influencing rates of pollen-tube growth, in theory the female flower also might hold considerable sway over which of the many pollen grains will fertilize her eggs.

Louisiana irises (in the genus *Iris*) are lovely, flowery occupants of wetlands in the southeastern United States. These species come in various heights, and have colorful flowers that range from a beautiful lavender to brick reds and yellows. At least four of these iris species have different microhabitat preferences, but they also hybridize occasionally in transitional settings. Given that these species can produce viable and at least partially fertile hybrids, much interest has centered on reproductive barriers that might maintain the species' separate identities, and on what exactly happens when these barriers break down. One possibility is that pollen competition strongly influences who fertilizes whom.

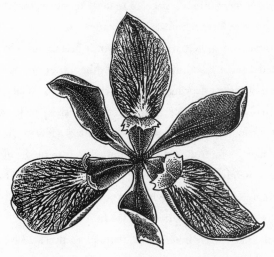

In greenhouses, scientific researchers conducted pollen competition experiments by jointly placing multiple pollen grains from various species on a female's stigma and later examining her progeny (using genetic markers) to see which of the males' gametes actually had fertilized the eggs. The researchers found that same-species pollen grains normally beat foreign pollen in these fertilization contests. The only exceptions came when foreign pollen grains were given a significant head start.

Pollen competition probably is important for Louisiana irises in the outdoors as well. In a perfectly fair contest, same-species pollen normally outrace different-species pollen to the eggs, but the races aren't always fair. Occasionally, a pollinating bee or hummingbird may deposit foreign pollen grains on a female's stigma before any same-species pollen has arrived. Only then is it likely that hybrid progeny will be produced.

Eating One's Own Kids

Cannibalism, or eating other members of one's own species, is a rather common phenomenon in the animal world and often seems to make a good deal of ecological and evolutionary sense to the lucky diner. By making a meal of one's compatriots, a cannibal not only may increase his own survival and reproductive prospects but simultaneously decrease the population of his potential competitors for food, space, or mates. Of course, cannibalism is a double-edged sword, for cannibals can be eaten too.

Cannibalism is common in many species of fish, some of which seem to have added a most puzzling twist to the phenomenon: parents apparently eat their own offspring as well. Such suspected "filial cannibalism" is particularly noticeable in nest-tending fish species, where a guardian male sometimes is observed gulping down a few of the eggs or fry from the nest he is tending. Filial cannibalism is to be distinguished carefully from "heterocannibalism" (the consumption of non-kin within your species), in part because the former behavior is far more puzzling evolutionarily. Why would a father fish eat his own babies?

Several possibilities have been raised. Perhaps natural selection favors

filial cannibalism when a nest-guarding male otherwise is faced with weakness or starvation. A truly dedicated father seldom may be able to leave his nest to forage, so to get through the tough nest-guarding times, his best or only option may be to sacrifice some of his progeny. Although clearly detrimental to those who are eaten, in the longer term such filial cannibalism may be beneficial or necessary to the male and his remaining kin in allowing at least some members of this and subsequent broods to survive.

A second possible explanation for filial cannibalism is that the eggs or children that a male eats from his nest are fungal-infected, or otherwise diseased, and would not survive anyway. Indeed, by eating any fungused eggs, a good-housekeeping father could stop the spread of infection within his nest. A third possibility is that filial cannibalism is simply a nonadaptive behavioral by-product of a fish's general voraciousness. In most other contexts, it clearly benefits males to be aggressive eaters. In this view, filial cannibalism is just a little mistake, a minor slip-up. In regard to the final tally of offspring produced, it matters little if a few babies are consumed from a nest that may contain dozens to hundreds of eggs and developing embryos.

Still, if they're going to engage in cannibalism, wouldn't it be wiser for fish to eat other males' offspring rather than their own, if at all possible? Well, perhaps they do. Recall (from the first essay in Part 5 and in "Egg Thievery and Nest Piracy," above) recent genetic discoveries that not all babies within some fish's nests are sired by the nest guardian. Such instances of "allopaternal care" can result from fertilization thievery (cuckoldry) by other males or from instances of either nest piracy or egg thievery by the resident male. Thus, many fry in a nest are tended by a foster father rather than their true sire. So, might most of the apparent instances of filial cannibalism instead really be heterocannibalism by foster males on their adopted offspring?

In a recent study, freshly eaten embryos were discovered in the stomachs of some cannibalistic tessellated darter *(Etheostoma olmstedi)* males who were tending their nests. From genetic paternity analyses conducted on nearly forty partially digested babies, it was proved that indeed they had been eaten by their respective biological fathers. Thus, the phenomenon of filial cannibalism, although seemingly odd, is real.

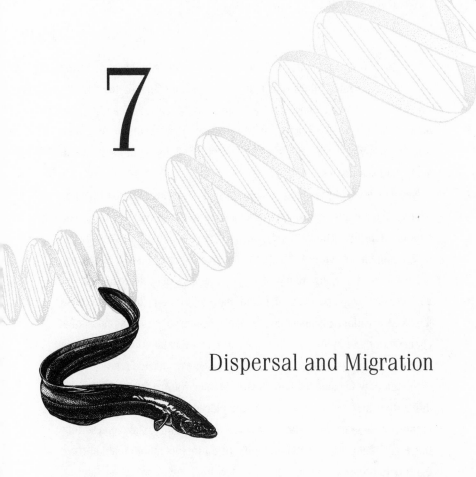

7

Dispersal and Migration

"D ispersal" is any spatial movement of an organism away from its birth site or prior breeding location. "Migration" refers to a periodic, often seasonal movement to and from a given geographical area, often along a consistent route. Wherever parents or their gametes or offspring move, naturally they take their genes with them, and this process is termed "gene flow."

Traditionally, organismal dispersal and migration have been studied by field naturalists who monitor the movements of tagged individuals. For example, professional birdbanders routinely capture birds and place metal or plastic rings on their legs. Later, if these color-coded or numbered bands are observed or recovered elsewhere, they can help to reveal a bird's dispersal behavior or migratory circuit. Similarly, by fitting animals with radio collars that transmit electronic signals to properly tuned receivers (which nowadays include space satellites), scientists can monitor even the most erratic or long-distance movements of otherwise difficult-to-track creatures such as mountain lions, penguins, tunas, seals, and sea turtles.

Nevertheless, physical tags have serious limitations as a source of information on animal movements. Because they do not transfer from parents to offspring, artificial tags can contribute little to an understanding of the cumulative, multigenerational consequences of dispersal and migration. Indeed, tagging studies of a given species can be misleading about such matters when patterns of animal movement today differ from those in past generations (as might be the case if ecological circumstances have changed, for example). Finally, physical tags are essentially impossible to place on tiny creatures such as many insects or marine larvae.

Fortunately for modern naturalists, Mother Nature foresaw all these difficulties. In the structure of DNA, she placed unique genetic tags on all creatures, no matter how large or small. Furthermore, she ensured that these molecular tags would be transmitted in specifiable fashion from parents to progeny, and she dictated that these tags were to accumulate intelligible differences (via mutation) across evolutionary time. Thus, properly read, molecular genetic markers can provide historical chronicles of when and where particular dispersal or colonization events took place. What more could scientists ask of nature's self-contained historical diaries?

The Tale of Mother Goose

During the mild summer months, snow geese *(Chen caerulescens)* congregate in large numbers at about a dozen major rookeries (nesting sites) in the tundra of extreme northern Canada, Alaska, and Russia. On the densely populated rookeries, these lovely birds of either white or bluish plumage lay their eggs in a crude nest on the ground, and each adult pair subsequently tends its brood of several goslings. Most of the nesting locations are separated by many hundreds of miles. Clearly, the species must have colonized these sites quite recently in evolutionary time, because even 15,000 years ago this land was buried under hundreds of feet of ice from the last great glacier of the Pleistocene geological epoch. At its peak, this vast continental mass of ice had pushed about as far south as today's Ohio and Missouri Rivers, well beyond where any of the snow geese colonies now reside.

Over the last few decades, field biologists have placed leg bands on hundreds of nesting snow geese, and the tag returns have revealed much about the modern-day travel habits of these waterfowl. Being sensible northerners, snow geese migrate south in the autumn to warmer climes, notably in California and Texas. Birds from multiple nesting sites mix on these wintering grounds, and this is where courtship and pair formation take place. Some mated pairs will remain together for years.

Come the spring, each dutiful male usually follows his mate back north to her natal rookery (the breeding colony where she hatched). As a logical consequence of this reproductive cycle, genes carried by males (more so than by females) should be exchanged between different rookeries in each goose generation. If so, the inter-rookery movement or flow of nuclear genes (dispersed by the mate-following males) could be far greater than that of the mtDNA genes that are transmitted only by females (who are far more faithful to their natal colonies).

Do the recently acquired genetic data actually match these hypotheses about population patterns in snow geese? The answer, which at first came as a great surprise, is "not entirely." Nuclear genes did prove to be well mixed among snow geese rookeries, as expected. However (and this was the eye-opener), the maternal mtDNA lineages were well mixed among rookeries also, even in the female birds.

The explanation for these counterintuitive genetic findings may lie in the different but complementary kinds of information provided by artificial birdbands versus nature's genetic tags. From the recoveries of leg bands, the fidelity of living females to their home rookeries was documented, but these data reveal nothing about the earlier history of the species or how its rookeries were colonized. The genetic data, by contrast, retained a record of historical gene movement within the species. Close matrilineal ties (as revealed by the mtDNA markers) indicated that either the modern rookeries were colonized in recent times from a shared ancestral stock or females had moved among the different colonies fairly recently, or both. The genetic data provide a historical view of geese dispersal that enriches and complements the kinds of information on current movements gleaned from the tagging studies.

It's important for scientists to remember that a full appreciation of dispersal and movement in any species requires an integration of historical and contemporary perspectives. As historians well know, the past can inform the present, and the present can be informative about the past.

The Fish That Braves the Bermuda Triangle In

most native fish species, populations in separate rivers or lakes display at least some genetic differences due to cumulative effects over time of natural selection, genetic drift (random genetic fluctuations in populations), and other evolutionary forces. Such geographic "population genetic structure" (population structure or genetic structure, for short) arises almost inevitably because fish are confined to water, and different water drainages by definition are disconnected from one another at the present time. Thus, any widely distributed species of fish that occupies freshwater streams yet doesn't show population genetic structure is of special scientific interest.

One such fish may be the American eel, *Anguilla rostrata*. This species, whose slimy adults look like foot-long sections of a garden hose, is abundant throughout North and Central America in brackish waters as well as freshwater streams draining into the Atlantic Ocean or the Gulf of Mexico. Traditionally, the American eel has been viewed as a likely candidate for breaking all the normal rules on population genetic structure. Why? Because the species has a peculiar "catadromous" life cycle. Catadromy is a lifestyle in which a fish spends most of its life in freshwater but then moves to the sea to spawn.

The eels that inhabit marshes and freshwater streams are juveniles and preadults. They arrived there following a long journey from the open sea that gave them birth, nurtured them for many months as translucent leaf-shaped larvae, and then distributed them to the shoreline habitats as slim, finger-long elvers. The females in particular then swam inland, up rivers, streams, and creeks, sometimes for many hundreds of miles. At about ten years of age, this cohort of eels nears sexual maturity, and the irresistible call of the sea beckons once again.

Leaving their adopted freshwater homes, the eels embark on a return journey of epic proportions, their destination none other than their birthplace, the Sargasso Sea, located southeast of Bermuda in the middle of the western tropical Atlantic Ocean. There, in the deep abyss, they probably spawn en masse and die, thereby completing one of nature's truly astounding life stories. This spawning, however, has never been witnessed directly by human eyes. Scientists capture baby eels in plankton nets trawled through the open ocean, and it is from these recoveries that the location and timing of the spawning event were deduced.

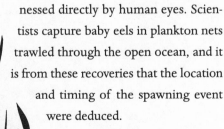

In the conventional wisdom, eels from throughout the Americas converge on this single area to spawn in one gigantic random-mating swarm, and the resulting larvae then swim and drift in ocean currents back to the continent. If so, the elvers entering any continental stream might all be random draws from that single mating pool. Accordingly, the peculiar life history of the eels led to the hypothesis that freshwater eel populations in North American streams might lack appreciable genetic structure.

Are these populations, indeed, genetically uniform? For the most part, the answer is yes. Based on molecular surveys, eel populations throughout eastern North America are nearly indistinguishable from one another genetically, a highly unusual finding for any freshwater fish but an outcome nonetheless consistent with the orthodox wisdom stemming from the peculiar life history of this species.

However, one genetic surprise was in store. A closely related species of eel *(Anguilla anguilla)* inhabits streams of Europe but also is thought to spawn in the Sargasso Sea. Might it too be a part of the same interbreeding melee? In genetic assays, European eels proved to be readily distinguishable from American eels, confirming that at least two highly distinctive spawning populations (and, therefore, species) truly are present in the North Atlantic region. Furthermore, there may be some slight genetic differences among eels from different European rivers as well. In any event, the baby European eels that are spawned in the Sargasso Sea survive an even longer journey. They drift for months or years across the entire North Atlantic Ocean before settling in European rivers, only to return years later, as adults, to spawn and die in that alluring region of the Bermuda Triangle.

Many mysteries remain. How does a tiny eel larva, who has never seen its parents, make its way across thousands of miles of open ocean to the proper continent, and how then does it find a suitable stream and adjust to freshwater life? When it matures a decade later, how does it know to swim downstream and then navigate a seemingly featureless sea to find spawning partners in the Sargasso Sea?

The Migratory Circuit
of a Whale

How would you like to spend your summers in cool southern Alaska and your winters surrounded by the balmy waters of the Hawaiian Islands? Many native Alaskans indeed do exercise this delightful travel option. One of these "snowbirds" is the humpback whale, *Megaptera novaeangliae*.

During the summer months, many humpbacks in the North Pacific feed and luxuriate in the plankton-rich waters off southeastern Alaska. As the sun falls lower in the autumn sky, these animals leave for their wintering home in Hawaiian waters, more than 2,000 miles away. There, females calve offspring who will migrate with their mothers back to Alaska the following spring.

Other humpback populations around the world have different seasonal migrations. For example, another group in the eastern Pacific that summers in central California winters primarily in western Mexico, and many humpbacks in the North Atlantic who summer off eastern Canada, Greenland, or Iceland take their winter holidays near the Dominican Republic. Unfortunately, in the last few centuries humans have hunted the humpback, like other species of great whales, to the brink of extinction.

Genetic assays have helped to answer several questions about the life history of humpback whales. For example, are geographic populations of this species genetically distinct, and if they are, what keeps them separate in this vast fluid environment? The first trick was to get small tissue samples for genetic analysis without bothering or harming the whales. This challenge was solved by the invention of a self-releasing biopsy dart that can be fired from a crossbow to obtain tiny bits of skin and blubber from free-swimming individuals. The second trick was to identify suitable genetic markers for this species in the laboratory. These proved to be abundant in mtDNA.

Once gathered and analyzed for many samples of humpback whales, the genetic data indicated that matrilineal differences between various regional populations are shallow (recent) evolutionarily, but nonetheless consistent. For example, closely related "female family names" that tended to characterize humpback whales in the Alaska-Hawaii migratory circuit differed from those shuttling annually between California and Mexico. Thus, within the North Pacific, there are at least two distinctive population stocks. This finding is not without conservation relevance, because if marine biologists are to help these wonderful animals survive into the future, they should know how many populations exist and where the whales reside and migrate.

Such spatial structure in an exceptionally large, mobile creature of the open ocean was not expected, and it must be the consequence of maternally directed fidelity to specific migratory destinations. When a female calf follows her mother northward, presumably she learns a migration route that she herself will follow for life. Decades later, she will teach this route to her children, just as her maternal grandmother taught it to her mother. Thus, in humpback whales (as in many other group-living cetaceans), social behavior plays a very important role in determining how populations are structured genetically across space. In other words, even in this most fluid of physical environments, the open ocean, extended whale families can be highly viscous.

A Salute to Salmon

When it comes to reproductive migrations, there's no accounting for fishes' varied tastes. Catadromous eels (see "The Fish That Braves the Bermuda Triangle," above) must find the open ocean irresistible, because upon reaching sexual maturity they leave the comforts of their freshwater streams to venture forth thousands of miles, returning to their birthsite in the open sea to spawn and die. "Anadromous" species, such as salmon, must have an equally compelling urge to move in the opposite direction. When a salmon reaches sexual maturity after several years in the ocean, it too feels the allure of its birthsite. Migrating to the coast, it enters a freshwater stream, battling its way inland perhaps for a thousand miles, over treacherous dams and waterfalls and through bruising shallows. Finally reaching the stream where it had hatched several years earlier, it spawns and, totally exhausted from the heroic effort, soon dies.

In its frigid spawning stream, a female salmon lays her eggs in a shallow depression (redd) dug into the gravel. Males then release their sperm, fertilizing her eggs, which will hatch within a few months. Each larva (alevin) remains buried in the gravel until its yolk sac is absorbed. Only then will it become free-swimming and start its long journey to the sea. Young salmon are well adapted to freshwater but soon begin a physiological and developmental transformation as they enter saltier waters. Once in the ocean, some salmon remain in coastal areas but others travel extensively. Several years later, each mature salmon returns to its home river to spawn.

The spawning phase of this migratory life cycle makes anadromous salmon highly vulnerable to human interference. Due to overfishing, dam construction, and habitat destruction, the spawning runs of many native salmon stocks along the Pacific coast of North America, for example, have been reduced by more than 90 percent in the last century. About half a dozen species of anadromous salmon in the Pacific Northwest (the chinook, sockeye, coho, chum, pink, steelhead, and sea-run cutthroat) are all now major objects of conservation concern.

From a genetic point of view, an interesting natural history question is whether the homing propensity of anadromous salmon has produced striking population genetic structures in the Pacific Northwest region. Depending on the species and area surveyed, diverse molecular genetic outcomes have been observed. In some cases, even adjacent streams or tributaries house genetically divergent salmon populations, indicating that natal homing (the propensity to return to birthsite) across the generations has shaped these species' genetic structures profoundly. In other cases, notably along recently de-glaciated coastlines, population structure has proved to be less pronounced. This confirms (as must have been the case) that these newly opened streams were colonized recently in evolutionary time by fish who had not returned precisely to their natal nurseries.

Tragically, most native salmon stocks in the Pacific region are now in severe jeopardy, despite costly and often desperate measures to save these animals. Fish ladders have been constructed around dams, draconian fishing regulations have been put into effect, and rivers have been stocked with huge numbers of hatchery-reared fish. Much of this effort has been to little avail, and some of the fishery practices may even be counterproductive. For example, careful studies have revealed that native salmon in their natural habitats have evolved special adaptations, such as an innate (instinctual) "knowledge" of the proper compass head-ing for migration. As these native fish are replaced by fat-and-pampered salmon born and raised in hatcheries, will these magnificent species soon lose their finely honed instincts for migration and survival in the wild, and with it the incredible wisdom of their ancestors?

The King of Migration
The migratory capabilities of many of nature's creatures are truly astounding. How an eel, salmon, whale, or bird can find its way around the globe, sometimes navigating thousands of miles to a particular wintering or breeding destination it may never before have seen, nearly defies human comprehension. Yet in some ways, all these accomplishments are trumped by the performance of an invertebrate animal, a roving king among princely nomads, the monarch butterfly *(Danaus plexippus)*.

This aristocratic wanderer, dressed in orange and black, is a familiar sight across North America during the summer. However, belonging to a tropical group of butterflies, these insects are intolerant of freezing temperatures, and in the autumn they migrate southward in massive numbers. The wintering destination for all monarchs east of the Rocky Mountains is an isolated pocket of mountain forest in central Mexico, about seventy-five miles from Mexico City. Monarchs west of the Rockies gather on the central California coast. At these two locations during the winter, hundreds of thousands of resting adults literally drape the trees like so many orange-dappled leaves.

These navigational and exertional feats are remarkable enough, but even more incredible is the fact that each several-thousand-mile journey is multigenerational. Except for the overwintering adults, each individual lives only about two months, so the monarchs returning to Mexico or California each autumn are the fourth- or fifth-generation descendants of the monarchs that ventured forth from these wintering sites the prior spring.

The process begins when each monarch travels northward in early March, laying her eggs on favored food plants (notably milkweed) along the way. The winged vagabond dies by April, however, so the journey is continued by her offspring, who within a span of a few weeks have hatched, grown through the caterpillar stage, pupated, and emerged from bright emerald cocoons to wing their way farther northward. They too mate and lay eggs as they travel but then die by June. Like members of a coordinated relay team, their children, grandchildren, and great-grandchildren successively grab the migrational baton, finally reaching the northern United States and Canada before returning southward in the fall.

The forebears of monarchs must at one time have belonged to a single ancestral population that somehow became sundered into the eastern and western units seen today that have separate migratory routes and wintering areas. How long ago did this separation take place, and how many genetic differences have accumulated between these two groups in the interim? Molecular markers thus far have failed to detect appreciable genetic differences between the eastern and western monarch forms. Most likely, then, the population separation was a relatively recent event, perhaps postdating the retreat of the last great continental glacier a mere 15,000 years ago.

This genetic finding also means that monarch butterflies must be able to alter their migratory pathways quite quickly in evolutionary time. Further support for this inference comes from the observation that monarchs, which are native to the New World, spread across much of the globe in the 1800s (probably with human assistance). They now occupy Australia, New Zealand, and many South Pacific islands, for example.

Monarch butterflies are of special concern to conservation biologists and nature lovers for several reasons: they are a regally beautiful part of the natural world, their migrational behavior is uniquely fascinating, and their populations are vulnerable to extinction because they are confined to such small wintering areas. Even a few bad winter freezes, forest fires, or disease epidemics, for example, could wipe out nearly all the monarchs in North America.

The genetic findings in this case do not necessarily get to the real heart of the mysterious navigational issues involved. Nonetheless, almost any uncovered secrets about these butterflies can be precious.

Green Turtle Odysseys

Green turtles *(Chelonia mydas)* are dedicated mariners, eating, sleeping, and mating in their oceanic home. The only interlude from this seafaring existence is when a gravid female comes ashore to lay eggs. With deep sighs and tearful eyes, the several-hundred-pound beast hauls herself onto the beach, like some prehistoric monster from the deep. Using her flippers, she digs a hole into which she deposits about 100 eggs that look and feel like slimy Ping-Pong balls. Covering these over with sand, she then crawls back to the welcoming sea. If all goes well, her babies hatch in about eight weeks. Like the mother they will never know, the two-inch hatchlings traverse the beach and plunge into the ocean, where they will move vast distances and grow to maturity in about twenty years. The life cycle comes full circle when a mature female mates at sea and then returns to land to lay her eggs. Each female repeats this procedure several times per season on a two- or three-year interval throughout her adult life of fifty years or more.

Scientists take advantage of these brief turtle sojourns ashore by applying metal or plastic tags to the animal's flippers. Tag recoveries months or years later have revealed much about the migratory habits of these seafarers. For example, green turtles tagged at a nesting colony in Costa Rica later have been found thousands of miles away on marine feeding pastures in Florida,

Surinam, and the Lesser Antilles. However, each Costa Rican female normally returns to Costa Rica to nest throughout her adult life, regardless of where she wanders in the interim. The same can be said for females nesting at each of the green turtle's handful of major rookery sites around the world.

What remained unknown from tagging studies is whether the nesting beach to which a female is faithful as an adult is also her natal site (where she herself was hatched). This is because no physical tag has been devised that when applied to a tiny hatchling weighing only a few ounces can withstand prolonged exposure to the caustic marine environment as the youngster slowly grows into a quarter-ton adult. However, Mother Nature has applied tags to these turtles too, and geneticists now can read these DNA markers critically to address the issue of whether females usually return to their natal sites to nest (that is, whether "natal homing" is the norm).

Because mtDNA markers are maternally inherited, they are ideal for this problem. When green turtle rookeries from around the world were examined genetically, it turned out that almost every colony had its own unique suite of female "family names." This genetic finding constitutes strong evidence for natal homing, because, if females instead frequently nested at sites other than where they had hatched, over time the maternal lineages in the species would have become thoroughly scrambled among the different rookery sites.

Additional genetic discoveries were of particular interest in one unique geographic setting. Ascension Island, only five miles in diameter, is an isolated dot of land lying near the equator in the southern Atlantic Ocean, about 1,400 miles east of northern South America and 1,900 miles west of Africa. Ascension's landscape is stark, moonlike, and nearly vegetation free except near the summit of the island's central volcanic peak. Rocky shores are interrupted by a few stretches of sandy beach. Several seabird species nest on Ascension, but it is home to precious few land animals, most notably centipedes, scorpions, and some insects who somehow managed to colonize this lonely garrison.

Yet, each winter and spring, Ascension becomes the site of a spectacular biological event as ponderous green turtles emerge one by one from the pounding surf. Ascension's females arrive there following a journey from South America taking several weeks. The Brazilian coast offers extensive beds of turtle grass that provided these females with their most recent meal, and it is to these shallow marine pastures that these reptilian giants soon will return. This 2,800-mile round-trip odyssey must entail incredible navigational skills and astonishing feats of physiological endurance. So, why do Ascension Island nesters take on such daunting challenges when suitable nesting beaches in South America would be far easier to reach?

One intriguing hypothesis raised in the scientific literature is that some green turtles may have begun nesting on a proto-Ascension Island about 80 million years ago, when South America and Africa are known to have been separated only by a narrow sea channel. Over the ensuing tens of millions of years, due to plate-tectonic movements of Earth's crust, the continents slowly drifted apart to assume their current configuration. With regard to the turtles, the hypothesis is that females in each successive generation gradually lengthened their migratory circuit inch by inch as Ascension Island slowly moved away from the South American coast. If this speculative notion is correct, it would mean that turtle matrilines (female lineages) on Ascension Island today should be greatly different, genetically, from those of other modern-day green turtle colonies in the Americas or Africa.

This was an imaginative idea, but what do the turtles' genes actually have to say about the matter? It turns out that Ascension's green turtles show only negligible genetic differences from members of the same species that nest in South America and Africa. These findings strongly indicate that Ascension was colonized recently, probably within the last few thousand years, by gravid waifs hatched elsewhere. Although the ancient-colonization scenario for Ascension Island was both evocative and alluring, it proved to be wrong. This is the nature of science, which merely strives to tell it like it is and not how we might have dreamed it to be.

Sweet Bees with a Nasty Disposition

American honeybees *(Apis mellifera)*, so ubiquitous in the New World, are actually of Old World ancestry. The species is native to the Middle East, Africa, and Europe, but in recent times has been transported worldwide by beekeepers. First to arrive in the Americas came the Spanish, Italian, and northern European races, all introduced by humans in recent centuries for the pollination services these colonial insects provide and for their sweet-tasting honey. For a long time, everything seemed to be fine between these domesticated honeybee immigrants and their human keepers.

Then, in the late 1950s, honeybees from Africa were introduced into Brazil. In a process referred to as "Africanization," aggressive "killer bees" (of the popular press) soon spread rapidly in the New World, storming through South and Central America, Mexico, and recently reaching the southwestern United States. Unlike their mellower European counterparts, Africanized bees have a nasty temper and sometimes badly sting unlucky passers-by.

To understand what might have happened in the Africanization process, some background about honeybee lifestyles is necessary. A honeybee colony is an expanded family unit commonly headed by a single mother queen. In her first two weeks as an adult, this young princess and heir to the throne flies off from her home colony to mate with several drones (males) from other hives (see "Mating Champions of the Insect World," in Part II, on page 177). Using stored sperm from this hectic flight, throughout her life she then produces sterile female workers to tend her own colony. She also produces fertile daughters (some, like her, destined to become queens) as well as princely sons, using, respectively, fertilized and unfertilized eggs (see "Extreme Social Behavior and Gender Control," in Part 4).

Based on this life cycle, two distinct hypotheses were raised for the rapid spread of killer bees in the Americas. The first supposes that shortly after their arrival in the New World, African drones sometimes traveled great distances on their nuptial flights to mate with virgin princesses of European descent who already were resident on the continent. Hybridization

ensued, producing hon-
eybee offspring with Eu-
ropean mothers, African
fathers, and nasty killer-
bee attitudes. Alternatively,
the range expansion of killer
bees might have resulted from
"colony swarming," wherein
recently introduced queens of
African descent (and their faithful workers) sometimes abandoned their original hives and
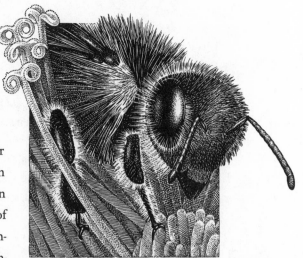
flew off long distances to establish new outposts in their New World
home. In that case, hybridization between African and American honeybees would not necessarily have been involved in the rapid expansion of
the killer-bee range.

Which hypothesis is correct? Molecular markers have provided some
answers. First, genetic surveys were conducted of honeybees throughout
their native range in the Old World. These revealed clear differences between honeybee races in Europe and Africa. Then, these distinctive genetic markers were employed in turn to screen Africanized honeybee
colonies in the New World. Under the hybridization hypothesis, in which
far-traveling African drones are involved, colonies at or near the advanc-
ing front of the killer-bee range should display mtDNA genotypes of European descent. By contrast, under the colony-swarming hypothesis these
same killer-bee colonies should carry African-type mtDNA instead.

The empirical results are now in, and they clearly show the involve-
ment of colony swarming in the spread of killer bees. However, similar
analyses, based on other molecular markers that are passed to the next
generation via both queens and drones, indicate that hybridization between honeybees of European and African descent has been a part of
the Africanization process as well. Thus, both colony swarming and hybridization appear to have been involved in converting some of our formerly mellow honeybees into potential killers.

The Ballast Travelers

From distant shores they came, interlopers without a name
In a ship, deep inside, they had hitched a long, hidden ride
As larval vagabonds, traveling so greatly beyond
Their former home.
Their journey 'cross the sea, an inadvertent odyssey
Lasted for many days, during which time they whiled away
The hours, swimming cold, in the vessel's watery hold
In full darkness.
Then to burst on the scene, unannounced and unforeseen
Their havoc to wreak, on those most innocent and meek
'Till they themselves became, major players in the game
Of life and death.
What does destiny hold, for these wee innocents, yet bold
Who in juvenile throngs, survived marine passages long
To invade foreign realms, trusting all to fate, at the helms,
In their voyage?

Who are these foreign invaders? The organisms to which my ballad refers are two aggressive species of European green crabs: *Carcinus maenas,* native to the eastern Atlantic Ocean, and *C. aestuarii,* of the Mediterranean. Nearly indistinguishable in appearance, these animals scurry about seawalls, jetties, and tide pools. These common shoreline species have colonized many regions around the world in the past century, often becoming conspicuous and disruptive players in their new environments.

For each invasion event, scientists have employed genetic markers to identify the particular cryptic species of crab involved and, thereby, to determine the geographic site of origin of the immigrants. For example, crabs that colonized New England, California, and Tasmania proved to have arrived from the eastern Atlantic, whereas those now settled in Japan and South Africa apparently emigrated on multiple occasions from both the eastern Atlantic and the Mediterranean Sea.

These green crabs constitute but a tiny fraction of the vast array of biological invaders that nowadays hitch intercontinental rides, often as tiny babies, in the ballast waters of ships' holding tanks. For example, re-

cent scientific studies have documented hundreds of nonnative species in North American estuaries such as Chesapeake Bay and San Francisco Bay. These unwanted intruders can cause considerable environmental trouble by altering natural ecosystems and sometimes forcing the extinction of species native to the region.

In the marine realm as elsewhere, exotic species are becoming ever more serious pests. When people move these creatures from ocean to ocean, intentionally or otherwise, the all-too-common result is that ecological balances are disturbed and native species suffer. It's like playing ecological roulette. Indeed, such introductions pose threats to global biodiversity that are perhaps second in overall effect only to direct human desecration of the environment.

Introduced species may cause serious economic problems as well. For example, zebra mussels *(Dreussea polymorpha)* in the Great Lakes of North America now occur at some locations in such high densities (up to 40,000 per square foot) that these animals clog intake lines for city water supplies and in general make huge nuisances of themselves. Yet this species was native to Eurasia. Introduced into the United States in about 1985, carried in the ballast water of a transoceanic ship, it has been a major pest in the Great Lakes region ever since.

Conventional wisdom holds that "successful" invasive species share several characteristics, such as wide tolerances for variable habitats, high reproductive rates, and high mobility. Unfortunately, it is these same kinds of biological attributes that tend to make invaders tough on native species and hard to control once they become established in a new area. For highly invasive species, an ounce of prevention is worth far more than a pound of cure.

In the case of green crabs, these belligerent little animals reproduce with zeal and quickly overrun their new settlements. They have a voracious appetite for clams, mussels, and sea urchins, and this brings them into competition with coastal fishermen as well as their fellow invertebrates.

8

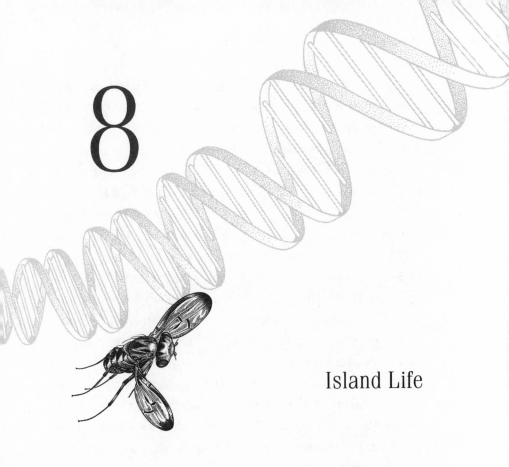

Island Life

Island plants and animals have had major impacts on the ecological and evolutionary sciences. When Charles Darwin first scouted the Galápagos Islands nearly 200 years ago and viewed their diverse creatures, compelling questions arose. Why were so many unique species of finches present in such a small land area? Why did each island have its own recognizable form of land tortoise? Where did the ancestors of these and other island species come from, and when? By pondering such questions, Darwin eventually came to challenge traditional explanations based on divine intervention, and his conclusions about natural (as opposed to supernatural) forces were forever to revolutionize all areas of human inquiry into the living world.

Island life continues to fascinate ecologists and evolutionary biologists. The creatures inhabiting archipelagos often have "adaptively radiated" into dozens and sometimes hundreds of closely related species that have worked out various ecological and behavioral ways to live together on their cramped island quarters. The physical proximity yet separation of islands within an archipelago seems to provide just the right blend of insularity, ecological novelty, and occasional opportunities for colonization to promote exceptionally rapid patterns of species' proliferation.

Sadly, native species on isolated islands also tend to be fragile, all-too-easily driven to extinction when humans alter their original habitats. Many island species have evolved special adaptations for peculiar ways of life in their ecologically unique homes. But because many oceanic islands formerly were free of terrestrial predators and some pathogens (disease-causing organisms), their inhabitants often lost the instincts and adaptations that otherwise might have enabled them to cope more successfully with rats, dogs, mosquito-borne diseases, and other aggressive exotics introduced in the last few centuries by humans. Finally, because most islands are relatively small, so too are the population sizes of the creatures who live on them.

For all these reasons, disproportionate numbers of island species have become extinct in recent times, following the arrival of human beings. For example, among the dozens of avian species that became extinct in the last three centuries around the world, more than 80 percent were "endemic" (native and confined) to islands. Furthermore, among native island species still surviving, sobering numbers are endangered or seriously threatened. Thus, island animals (and plants) are also of special concern to conservation biologists.

Darwin's Galápagos Finches

In December 1831, a twenty-two-year-old biologist named Charles Darwin set sail from Great Britain on what would be a five-year voyage aboard HMS *Beagle*. As the ship's official naturalist, Darwin's charge was to study and catalogue exotic plants and animals that the expedition might encounter on its westward, circumglobal journey. True to task, Darwin carefully described the diverse floras and faunas that he encountered at each of the Beagle's ports of call.

In South America, young Darwin marveled at the adaptations of plants and animals to environments ranging from the heights of the Andes to the depths of Patagonian valleys, and from Argentinean grasslands to Brazilian jungles. He pondered why the organisms differed from site to site and, even more so, from those of Europe. He puzzled over the meaning of long-extinct marine creatures preserved as fossils in sedimentary rocks, sometimes even recovered now on mountaintops gradually uplifted over the ages. But nowhere were Darwin's views on evolution shaped more profoundly than on a small island archipelago about 600 miles west of mainland Ecuador. Among the many creatures collected by Darwin on these Galápagos Islands were more than a dozen unique species of finchlike birds.

More than a century later, the evolutionary biologist David Lack (who actually studied these birds in far more detail than Darwin) named these species Darwin's finches. They are mostly a drab gray-to-black lot, but their diverse lifestyles more than compensate for what they lack in plumage variety. One species eats green leaves, another picks at the wounds of other animals and drinks their blood, and another perches on the backs of iguanas and rids them of ticks. Two species employ cactus spines or other sharp twigs as tools to extract grubs from hidden crevices. Other species of Galápagos finch specialize in foods ranging from soft insects to hard seeds. The size and shape of the birds' beaks vary accordingly, from slender tweezers to conical vices. These structural differences are exaggerated when two or more finch species co-occur on particular islands, apparently because natural selection has promoted species' adaptations that partition the available food resources.

Biologists long have realized that all the Galápagos finch species are one another's evolutionary relatives, but without explicit genetic data they couldn't be sure of the precise history of this taxonomic assemblage. What was the approximate time frame of this remarkable "adaptive radiation," and who is related most closely to whom?

Recent molecular assays of proteins and DNA have revealed that Darwin's finches are similar genetically, implying that most of the speciation events occurred within the last three million years. This makes sense, because this is the approximate geological age of the most ancient of the modern-day Galápagos Islands. Thus, the genetic data indicate that the radiation of finch species in the Galápagos Islands was explosive and quite recent, and furthermore that it followed a single colonization of the archipelago by an ancestral warblerlike finch from mainland South America. Much as Darwin first deduced and Lack and others confirmed, strong selection pressures, in conjunction with the splendid isolation of this desolate archipelago, conspired to create by natural means what otherwise might have seemed to be the direct but puzzling workmanship of a sentient god.

The research also yielded one modern surprise. In sensitive molecular assays, genetic lineages were found to be shared by several of the ground finch species (genus *Geospiza*), a subset of the Galápagos finch complex. Most likely, genes have been moved about from one species to another via ongoing hybridization. This genetic evidence confirmed field observations indicating that some of the ground finch species occasionally mate with one another and produce viable offspring that can "backcross" to the parental species. These findings in turn have called into question precisely how many distinct biological species of ground finch actually inhabit the archipelago.

Darwin suspected that the Galápagos finches (as well as some other avian groups in the archipelago, such as mockingbirds) originally diversified in response to natural selection, which led to the specialization of ancestral populations on different foods. In the twentieth century, several research groups confirmed these suspicions through critical observational and experimental tests. They also identified other relevant forces, including the isolating effects of disjunct island habitats, and "founder effects" (rapid evolutionary changes associated with small numbers of colonizers). Now, nearly 200 years after Darwin first visited the islands, scientists finally have an explicit genealogy for the Galápagos finches. Furthermore, the genetic data indicate that hybridization must be added to the list of evolutionary factors at play within the archipelago.

Whether or not Darwin's finches all are deemed separate species, interbreeding among some of them apparently contributes to the genetic variation upon which natural selection operates. Given Darwin's boundless intellectual curiosity, he no doubt would have been fascinated by this most recent discovery about the little brown birds that bear his name.

Beautiful Flies

Beautiful Flies The volcanic Hawaiian Islands have been likened to a conveyor belt in a geological factory dedicated to the production of unique land species. In a recurring theme, particular ancestral taxonomic groups that somehow colonized these isolated islands from distant mainland sources have adaptively radiated over evolutionary time into a plethora of species found nowhere else but on these beautiful isles of the central Pacific Ocean.

A case in point involves the approximately 900 species of fruit flies (family Drosophilidae) native to the Hawaiian archipelago. These curious islanders are distant relatives of the pesty little forms that materialize at ripe fruit in your kitchen. However, in their idyllic island homes, some of these flies have evolved into much larger, lovely creatures (called the "picture-wings") with elaborate courtship displays and pleasing patterns of wing veins. Although the Hawaiian Islands account for less that 0.01 percent of the earth's total land area, more than 25 percent of all the world's drosophilid species occur only in the aloha state. All these endemics appear to have stemmed from just one or two ancestral species that settled on the islands (from unknown continental sources) in the distant past.

The isolated Hawaiian archipelago itself was produced as the Pacific Plate, a portion of the earth's outer crust, moved over a stationary hotspot in the planet's mantle that periodically extrudes magma. As each volcanic island popped above sea level, over geological time it was transported slowly toward the northwest by the earth's crustal movements, gradually eroding and subsiding on its extended journey. The youngest isle is Hawaii itself (the Big Island), and the oldest of the current major islands is Kauai, about 250 miles to the north and west. Kauai was formed about 5.1 million years ago, and, hence, has moved "downstream" from the hotspot at an average rate of approximately three inches per year.

It would seem, therefore, that the tremendous proliferation of Hawaiian drosophilid flies occurred within just the last five million years. This assumption, however, is wrong. Several sets of molecular genetic data indicate that many Hawaiian drosophilid lineages diversified much longer ago than that; some of the inferred speciation events took place as much as 30 million years ago. How can this be, given the young geological ages of the modern isles?

As each newborn island migrates off the hotspot, it begins to decay as the ocean beats upon its shores, and winds and rain erode its hillsides. Each island's inevitable fate across geological time is to shrink slowly, eventually to become a low coral atoll, and finally to die by drowning. Thereafter, it is entombed as an underwater seamount. Such atolls and seamounts extend far to the northwest of the present-day Hawaiian Islands. These are all that remain of earlier isles in the Hawaiian chain, and they date back more than 40 million years.

These geological facts provide a key to interpreting the recent genetic findings indicating a relatively ancient origin for some of the Hawaiian fruitfly lineages. Many of the speciation events probably took place on ancient isles survived today only by atolls and seamounts. After each new fruitfly species arose on one of these archaic islands, if successful some of its descendants must then have island-hopped to one or more newly arisen islands in the Hawaiian chain. Thereby the evolutionary process of colonization and divergence by which the drosophilid species proliferated in the archipelago was continued. Judging by the large number of living fruitfly species and the considerable ages of some of their genetic lineages in the Hawaiian Isles, this geological conveyor belt and biological factory have been in successful operation for a very long time.

More Exotic Beauties of Tropical Isles

The Hawaiian archipelago also has produced the most spectacular variety of land birds ever found on oceanic islands: the honeycreepers (subfamily Drepanidinae). More than fifty distinct species of these lovely creatures apparently evolved from a single ancestor who long ago colonized these remote dots of volcanic land. Many of the species have strikingly beautiful plumages ranging in color from wine reds and golden yellows to jet black. The honeycreeper species that are still alive today usually inhabit only one or a few islands in the Hawaiian chain, typically in small numbers.

The evolutionary radiation of Hawaiian honeycreepers resulted in a great variety of adaptations, particularly with regard to the birds' diverse feeding habits. There are granivorous (grain-eating) and frugivorous (fruit-eating) species with triangular bills well designed to munch on foods ranging from hard seeds to fleshy pulps. There are nectivorous (nectar-feeding) species with long curved bills for sipping beverage from deep within a flower. There are also many insectivorous (insect-eating) species. Some of the latter possess acute bills, like tweezers, for gleaning tiny critters off leaves and other flat surfaces. Others have long or curved beaks, like dentists' probes, for poking and prodding insects from deep crevices. Still others have bizarre specializations, such as laterally asymmetric bills for prying open leaf buds that contain grubs.

This great array of adaptations raises a compelling question: How old, evolutionarily, is the Hawaiian honeycreeper assemblage? Based on recently gathered molecular genetic data, most of the Hawaiian honeycreeper lineages apparently separated from one another within about the last five million years, that is, within the geological time frame encompassed by the volcanic birth of the major islands that still are above sea level today.

What could account for such a rapid diversification of honeycreeper species? Islands in the archipelago are separated by long-standing ocean channels that undoubtedly limited the opportunities for interbreeding between honeycreepers on different isles, and thereby enhanced the likelihood that the isolated populations would diverge into new species. Yet, the islands are not so far apart as to preclude the occasional movement of small numbers of birds from one island to another, and this process could lead to additional rounds of population isolation and speciation. Furthermore, each island when colonized initially by an ancestral honeycreeper species provided an open ecological setting, relatively free of the intense competition that characterizes more mature habitats. Diversification among the avian populations on different islands then ensued as the newcomers specialized on varied food resources in their new homes.

If you travel to the Hawaiian Islands today, don't expect to be greeted by more than a small handful of these fascinating honeycreeper types.

In recent times (especially in the last two centuries), humans have caused the extinction of more than half of the original honeycreeper species, and most of the survivors are threatened or highly endangered. The honeycreepers' demise is attributable to the customary list of factors: deforestation associated with urbanization and agriculture; overhunting, especially for feathers used as ornaments; the proliferation of nonnative predators such as cats and rats; and human-mediated introductions to Hawaii of novel avian diseases (notably bird malaria, carried by mosquitoes). Tragically, what took nature millions of years to accomplish has been undone all too quickly.

Radiant Plants of the Hawaiian Islands

The Hawaiian archipelago also is home to a beautiful and exotic variety of native plants. Species with peculiar adaptations can be found in all the diverse natural habitats of the islands, including parched deserts and wet scrub, coastal dunes and bluffs, montane ridges, rainforests, dry woodlands, and soggy bogs. Several species even have gained toeholds in the cracks and crevices of barren lava flows and stark cinder fields from recent volcanic eruptions. Many plant species are confined to one or two of the main islands, apparently not having found their way across the intervening sea channels. What follows are in-

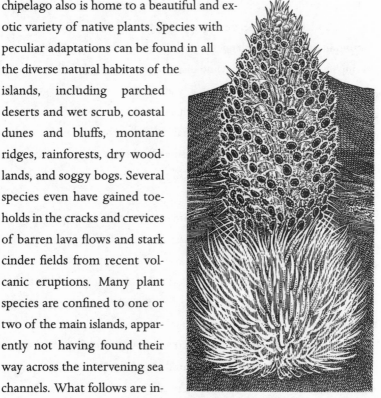

troductory snapshots of a few members of one particular endemic Hawaiian group known as the silversword alliance.

Argyroxiphium sandwicense is a rosette plant (leaves radiating from a single base) confined to dry scrub areas and lava fields of Hawaii and Maui. Like some large, spherical porcupine, it has stiff silvery leaves up to fifteen inches long that point outward in an intimidating ball. In a grand reproductive gesture, this shrub flowers only once in its life, and then dies. By contrast, another rosette shrub, *A. grayanum,* is confined to bog habitats on the island of Maui and normally bears fruit multiple times during its life.

Wilkesia gymnoxiphium is another shrub that flowers only once and then dies. Confined to dry habitats on Kauai, it too has rosettes of leaves fifteen inches or more long, but rather than radiating from a single ground point, they are borne on woody stalks shooting skyward a dozen feet high.

Dubautia ciliolata is a nonrosette shrub that grows on older lava flows, being found, for example, on the upper slopes of Mauna Loa, a volcanic mountain on the main island, Hawaii. This conventional looking shrub has inch-long leaves and tiny wind-dispersed seeds that blow far and wide across unsuitable habitat. Remarkably, another member of the genus, *D. latifolia,* is a woody vine, or liana. This vine climbs into the canopies of large trees in the wet forests of Kauai. Other species of *Dubautia* have varied growth forms, ranging from small bushes to treelike specimens. Leaf shapes in these species can be long and pointed, squat and oval, or anything in between.

Altogether, the Hawaiian Islands are home to five living species of *Argyroxiphium,* two of *Wilkesia,* and twenty-one of *Dubautia.* Apart from being prevalent endemic plants of Hawaii, what else do these silversword species have in common? Recently acquired molecular genetic data have provided an amazing answer. Despite their varied outward appearances, ranging from stout mats and cushion plants to shrubs, trees, and vines, *all* these silversword species evolved from a single common ancestor, a species of California tarweed (subfamily Mediinae) that colonized the Hawaiian Islands in the not-too-distant past.

Following that initial colonization, the genetic and morphological data point to a subsequent adaptive radiation of lineages leading to the astounding diversity of plant forms and lifestyles present in the Hawaiian silversword group today. Indeed, the diversity of growth habits among the Hawaiian silverswords greatly surpasses the total variation observed among all species of tarweeds in North America. Once the islands were reached, it was as if an evolutionary explosion took place within a group of plants that otherwise had been thwarted, on the crowded mainland, from reaching its latent potential for ecological diversification.

Thus, much like the leafy rosettes displayed by several silversword species, the evolutionary tree for the Hawaiian silversword alliance is like a radiant bush that literally burst rapidly upon the scene from a single basal stem.

The Clinging Goby Fish

On the windward, northeastern hillsides of the Hawaiian Islands, where rainfall can exceed 200 inches per year, several small torrential streams cascade down steep luxuriant slopes and into the sea. In these unlikely settings, thousands of miles in any direction from the nearest freshwater habitats on other islands or on continental landmasses, a small species of goby fish *(Lentipes concolor)* found nowhere else on Earth literally hangs on for dear life, in more ways than one.

In waters tumbling to the ocean from elevations greater than 2,000 feet, these fish cling tenaciously to submerged rocks by using a pair of modified pelvic fins that are fused into a sucker-like disk. These accomplished piscine rock climbers even scale their way up and over hundred-foot-high waterfalls. Although the individual fish obviously are hardy, the species is at risk of extinction due to its narrow geographical distribution and its special ecological requirements.

In truth, however, these gobies are not confined strictly to salt-free waters. Although the adults live exclusively in rushing freshwater streams, and spawn there, the hatched larvae swim out to the salty sea for a several-week cruise in the marine plankton. Then they return to babbling brooks to produce more of their kind and live out their remaining time.

This two-phase life cycle, spent largely in freshwater but including a brief seaward voyage, is termed "amphidromy." It differs from anadromy (typical of many salmon; see "A Salute to Salmon," in Part 7) in that the marine phase is strictly larval and of short duration. It differs from catadromy (characteristic of American eels; see "The Fish That Braves the Bermuda Triangle, in Part 7) in the same way and also in the fact that spawning takes place in streams rather than in the ocean.

Does the marine larval stage of gobies result in extensive gene flow (genetic exchange) among stream populations on different Hawaiian islands? Or, does each larva normally return to live and spawn in its own natal creek (where it had hatched)? It's quite impossible to tell simply from field observations or from the external appearance of the fish, but Mother Nature's internal genetic tags have provided the answer. By surveying molecular markers, researchers discovered that freshwater goby populations from the five main Hawaiian Islands are undifferentiated genetically. Results indicate that goby larvae must transfer between islands often, thereby causing considerable gene flow among the various stream populations throughout the archipelago.

Only five native fish species (all gobies or gobylike forms) inhabit Hawaiian streams, and each has its closest living relatives elsewhere in the Pacific region. Thus, these species colonized the archipelago independently. Furthermore, all five species are amphidromous. This paucity of fish species in Hawaiian freshwater habitats contrasts dramatically with the impressive adaptive radiations in several groups of native land-based species such as drosophilid flies, honeycreeper birds, and silversword plants (see "Beautiful Flies," "More Exotic Beauties of Tropical Isles," and "Radiant Plants of the Hawaiian Islands," above). Apparently, the ease with which larval gobies can travel among islands of the Hawaiian chain has promoted high gene flow and thereby squelched the kinds of geographic isolation normally necessary for high speciation rates. Such has been one evolutionary consequence of the mobile lifestyle of these island vagabonds.

Fabulous, Fabled Frogs

Dart-poison frogs (family Dendrobatidae) of New World rainforests are among nature's most strikingly gorgeous creatures. These species come in a dazzling array of flamboyant colors and patterns: brilliant blues, greens, reds, oranges, or yellows, usually dappled with patches of jet black. This resplendence as well as the pleasing personality of these tiny diurnal (day-active) frogs make them popular objects in a sometimes illicit pet trade.

As these exquisite charmers hop about openly on the rainforest floor, their bright colors are a warning to potential predators: "Don't you dare eat me!" Small pores in the frog's skin secrete defensive poisons that can be fatal if ingested. The toxin secreted by one of the dart-poison species, appropriately named *Phyllobates terribilis,* is among the most lethal of concoctions to be found anywhere in nature, with each specimen containing enough venom to kill ten adult humans. The Chocó Indians in the lowlands of western Colombia take advantage of this killing potion by rubbing blowgun darts along the backs of the amphibians. Poison-tipped, these projectiles then become highly effective in bringing down monkeys and other small game.

Another hopping jewel of special interest lives in Central America, in the Bocas del Toro Archipelago off Panama's northern (Atlantic) coast. Each island population of the strawberry dart-poison frog *(Dendrobates pumilio)* displays a different garb, with epidermal motifs of red, green, yellow, or orange, interspersed with black. Do these conspicuously varied populations in the Panamanian islands truly belong to one species, and if so, why do they show such amazing variety in color patterns?

From recent molecular marker evidence, all island populations of the strawberry dart-poison frog in Panama are extremely close to one another genetically. This finding bolsters earlier conclusions, based on the frogs' nearly identical chromosomal structures, toxic compounds, and mating calls, that these diversely dressed forms all shared a common ancestor in very recent evolutionary time. Also, the genetic findings generally are consistent with the conventional placement of all these brightly colored forms in the same taxonomic species.

Presumably, any of the eye-catching skin colors in these frogs would serve perfectly well in their primary role as brazen warnings to would-be predators, particularly those with color vision, such as birds. In toxic or venomous species of all sorts, ranging from insects to snakes, brilliant "aposematic colorations" (warning signals) are a common product of the evolutionary process. So, why then do strawberry dart-poison frogs display such great variety in warning colors, when any one color motif might suffice?

Various lines of evidence suggest that the answer does not reside in different diets or in other ecological differences among frogs on adjacent islands. Instead, many researchers now suspect that the rapid evolutionary divergence in skin colors was mediated by a form of sexual selection stemming from the extreme pickiness of females regarding their mates.

In this species (unlike several other dart-poison frogs in related genera), females rather than males provide most of the parental care. Mothers carry newly hatched tadpoles to small pools of water and then return periodically to these pools to lay additional eggs for their offspring to eat. Especially in species with such strong maternal care, sexual selection theory predicts that females should be very particular about whom they mate with, because a mistake means that much reproductive effort is wasted. Thus, current speculation is that the frogs' outrageous costumes diversified rapidly under sexual selection, as females on separate islands began choosing mates with differing color patterns. Why were different colors preferred? This is a question for which there is as yet no answer.

One objection might be raised to the female-choice hypothesis—the bright colors are displayed by both males and females, unlike the case in most other species under strong sexual selection, where females are drab. However, as noted above, the brilliant colors in the dart-poison frogs serve primarily as warning devices to predators. Although most human observers (and, presumably, the dart-poison frogs themselves) perceive these flashy skin colors as irresistible, predators must find them exceptionally repulsive. Beauty truly is in the eyes of the beholder.

Caribbean Cruises More than 1,200 species of terrestrial vertebrate animals inhabit the West Indies—islands that include the Greater Antilles (Cuba, Jamaica, Puerto Rico, and Hispaniola), the Lesser Antilles, Bahamas, and peripheral islets. Many are birds and bats, whose ancestors simply flew to the islands from distant realms in North, Central, or South America. However, many species of tropical frogs, toads, lizards, snakes, and turtles also are prominent among the native islanders of the Caribbean region. How they arrived on the islands is a bit more puzzling.

Assuming that the non-flying animals didn't arise by spontaneous generation, and given that they can't hop, slither, or crawl across water, how in the world did these land-based creatures manage to colonize such distant islands? Clearly, their ancestors must have arrived following journeys from their native homelands on the adjoining continents, but exactly how and when the historical emigrations took place long remained a mystery.

Two competing hypotheses were advanced. Under the "vicariance" view, many West Indian endemic species are the descendants of ancestral taxa that colonized the islands in the late Cretaceous period (some 80 million years ago). In that ancient time, the islands were immediately adjacent to the continental mainlands, having just separated from them geologically. Thus, the emigrants' journeys would have been short and easy and would have taken place in a relatively brief span of time long ago. Alternatively, under the "dispersalist" scenario, the islands were colonized much later in geological time via rare extended passages over wide seas (perhaps when the animals floated for weeks or months on rafts of vegetation). In that case, each journey would have been long and hard and, collectively for the various species involved, would have occurred over a wide range of evolutionary times.

To determine which scenario—ancient vicariance or later dispersal—better accounts for the presence of terrestrial vertebrates in the West Indies, genetic comparisons (from protein and DNA assays) have been made between dozens of endemic island species and their nearest relatives on the continental mainlands. The idea has been to estimate the approximate evolutionary date of arrival of each taxonomic group on the islands, based on the observed genetic distances as interpreted under appropriately calibrated molecular clocks.

From the molecular genetic data, three generalities have emerged about the origin patterns of fauna in the region. First, dispersal over water at various times throughout the Cenozoic era (the most recent of the geological eras, beginning about 65 million years ago and extending to the present) accounts for the origin of most of the fauna currently inhabiting the West Indies. Thus, the disperalist hypothesis and not the vicariance scenario was supported. Second, most of the avian and bat lineages that colonized the islands came from North or Central America, which makes sense because their continental margins generally are closest to the West Indies. Third, most of the amphibian and reptilian lineages came instead from the more distant shores of South America.

The latter finding seems astonishing. How did creatures that can't fly or swim manage to traverse hundreds of miles of open Caribbean Sea? A compelling clue comes from the observable patterns of oceanic current. In the southern Caribbean, surface waters circulate primarily from southeast to northwest. This oceanic flow tends to transport rafts of vegetation (nature's flotsam and jetsam) from river mouths in eastern

South America toward the West Indies. Thus, if land-based animals occasionally hitched week-long rides on these floating mats (as they are known to do today), any fortunate survivors might have found themselves stranded on the shores of some exotic isle.

Many of the successful colonists thrived in their newfound island paradises. Indeed, the descendants of some immigrants settled on additional islands, often leading to further speciation within the West Indian Archipelago.

9

Species Proliferations

Spectacular adaptive radiations (explosive proliferations of re-lated species during the evolutionary process) are not necessar-ily confined to island archipelagos (Part 8). Many wonderful examples are evident in the evolution of flora and fauna on continental landmasses too, as well as in the open sea. For each such radiation, scientific topics commonly addressed concern the history of the speciation episodes and the genetic or ecological factors responsible for the rapid multiplication of kindred species. In short, how many species arose in each radiation, and when and how did the speciations take place? This section provides some thought-provoking examples of how such evolutionary questions have been tackled, using molecular genetic markers, in several especially intriguing groups of plants and animals.

Warbler Wardrobes

Eurasian ornithologists are jealous of those of us in the Americas who can enjoy at our doorstep the exceptional charm and diversity of the New World warblers (family Parulidae). More than fifty species of these energetic woodland pixies spend their summers in North America perpetuating their kind before departing for warmer climes in the late summer and fall. Each autumn, Latin American birders welcome the return of these northern migrants, who join numerous other warbler species that live year-round in the tropics. Conversely, each spring, birders across the United States and Canada anxiously await the migrational return of these songsters from their winter homes in Central or South America and the Caribbean.

Particularly during the breeding season, these avian jewels of the New World are a lovely lot indeed. In North America, there is the prothonotary warbler of southern swamps with its blazing-orange head and breast. There is the sweet-songed yellow warbler of willow-lined ponds, whose sun-bright breast is streaked with a lovely burnt red. There is the black-throated blue warbler of deep deciduous woods, with snow-white belly contrasting beautifully with jet-black throat and steel-blue back. There is the American redstart of the forest understory, which looks like a lively orange and black butterfly as it fans its tail and engages in aerial acrobatics while catching flying insects. There is the cerulean warbler of the highest treetops, which descends only rarely to give us a brief glimpse of its lovely sky-blue crown and back.

The list goes on. There is the sharply dressed black-and-white warbler that shows off its pin-stripe suit while spiraling around tree limbs in search of bark insects. The worm-eating warbler has a suit of olive and tan accented by a boldly striped head beret. Other species of New World warbler are more flamboyant in their dress, with spectacular wardrobes alluded to by their common names: the golden-winged warbler, blue-winged warbler, black-throated green warbler, golden-cheeked warbler, yellow-throated warbler, bay-breasted warbler, chestnut-sided warbler, hooded warbler, painted redstart, and dozens more.

Old World forests are inhabited by dozens of insectivorous warbler species too (family Sylviidae), but in stark contrast most of these are a dingy, somber lot. The Eurasian warblers are of dull yellow or brown motif, nearly indistinguishable from one another to the human eye. Their common names reflect this plainness: brown bush-warbler, oliva-ceous warbler, ashy-throated warbler, buff-browed warbler, plain leaf-warbler, sombre leaf-warbler, dusky warbler, smoky warbler, and grey-cheeked warbler. Many Old World warblers have lovely songs, however. Indeed, to tell these songsters apart, birders often rely more on their vo-calizations and habitat preferences than on their feather patterns.

Given their highly varied plumages, one might suspect that the New World warblers are an ancient evolutionary assemblage compared with their Old World counterparts. After all, it might take considerable evo-lutionary time to accumulate such pronounced differences in feather-color patterns. This intuitive hypothesis was wrong.

By genetic yardsticks, the New World warblers actually have proved to be *less* divergent from one another than is the case for their Old World fellows. Thus, according to this recent molecular evidence, the primary bursts of speciation in several groups of the ebullient New World war-blers occurred during the Pliocene or Pleistocene epoch (that is, within the last five million years), whereas the assayed taxonomic groups of drab Old World warblers apparently evolved into different species and diversified over a considerably longer period of evolutionary time be-ginning well before the Pliocene.

New World and Old World warblers both partition the habitat by be-havior more so than by adaptive differences in body features. Many war-bler species inhabiting the same geographic area ("sympatric" species) are close in size and shape and have similar bills, yet they live and for age differently. For example, five widely sympatric warbler species in the genus *Dendroica* normally feed in different tree sectors: low branches (myrtle warbler), inner limbs at modest height (bay-breasted warbler), inner and outer branches at moderate height (black-throated green warbler), high outer branches (blackburnian warbler), and tree crowns (Cape May warbler). Although these birds occur together in coniferous forests across the northeastern United States and Canada, they utilize different resources.

In any event, the variety of plumage patterns in New World versus Old World warblers has proved to be a poor predictor of the relative evolutionary ages of these two groups. Fashionable New World warblers have evolved avant-garde plumages comparatively rapidly, whereas the staid Old World warblers have stuck to conservative plumage themes for eons. Yet both groups have speciated prolifically, and most of their proteins and DNA sequences have gone on evolving at standard avian rates.

"Mosaic evolution" is a term that describes any such heterogeneity in the speed at which different kinds of organismal features (such as plumages and proteins) evolve within an evolutionary lineage. Thus, the genetic and observational analyses of warblers have contributed to a broader scientific lesson: behavioral, morphological, and molecular features can march at their own paces, apparently to the beats of different evolutionary drummers.

Unity and Diversity in the Winged Aussies

When European naturalists reached Australia a few centuries ago, they encountered many types of native birds that seemed quite familiar to them. There were fairy wrens, tiny avian sprites who, much like their ecological counterparts in Europe, have trilled songs, a pert attitude, and an oft-cocked tail. Thornbill species in Australia were reminiscent in behavior and appearance of the Old World warblers in English gardens. Australian sittellas, with their distinctive habit of spiraling down tree branches in search of bark insects, reminded the European naturalists of tree-circling nuthatches back home. And there were the Australian treecreepers, who approach their bark-probing tasks in just the opposite direction, slinking up tree trunks, similar to how brown creepers work the trees in England.

Understandably, European naturalists assigned the Australian birds to traditional taxonomic families where they seemed best to fit. For example, sittellas were put into Sittidae (the nuthatch family of the Northern Hemisphere), and Australian treecreepers were placed into Certhiidae (a taxonomic family of Eurasian and American creepers). Such was the state of affairs in Australian avian systematics until the early 1980s.

Then, to everyone's amazement, genetic data prompted a fundamental reevaluation of all these traditional notions. In molecular assays, many of Australia's native birds proved to be only distantly related to their ecological counterparts on other continents. Instead, Australian species as diverse as fairy wrens, thornbills, sittellas, and treecreepers seemed phylogenetically closer to one another than to species elsewhere in the world that share their respective lifestyles, appearances, and behaviors. In other words, the molecular data indicated that much of Australia's diverse avifauna had evolved from a common ancestor on that continent. Some other Australian specialties reportedly included in this ancient adaptive radiation were scrub-birds, lyrebirds, fantails, honeyeaters, and bowerbirds.

If this genetic interpretation is correct (it remains contentious in some scientific circles), then the traditional taxonomic assignments for Australian songbirds improperly summarized these birds' evolutionary relationships to their ecological analogues in Europe and North America.

More important, it also suggests that the evolutionary history of much of Australia's avifauna roughly parallels that of its mammals.

Biologists long have been impressed by the remarkable radiation of marsupial (pouched) mammals in Australia, which led to diverse forms roughly similar in appearance and lifestyle to various placental mammals on other continents. For example, herbivorous kangaroos are more or less the ecological equivalents of deer or antelope, wombats are like rodents or woodchucks, and the Tasmanian wolf (now extinct) filled the carnivorous role of wolves in the Northern Hemisphere. But, thanks to a brood pouch and other distinctive traits shared by kangaroos, wombats, Tasmanian wolves, and all other marsupials, systematists have always correctly understood that an evolutionary unity underlies the outward morphological diversity of these Australian specialties.

Now, if the genetic deductions are correct, a similar evolutionary situation applies to Australia's songbirds, many of which would appear to have arisen from a shared avian ancestor on the continent. For the last 60 million years, Australia has been the most secluded of Earth's major landmasses (inhospitable Antarctica excepted). In this isolated evolutionary setting, songbirds (just as much so as the marsupial mammals) flourished and diversified into varied niches, often evolving similar kinds of adaptations to those displayed by unrelated creatures in other parts of the world.

Evolutionary Trees and
Elephants' Trunks
Hyraxes (family Procaviidae) are chunky, compact mammals, about the size of a loaf of bread, that scurry about vegetation or boulder outcrops on the savannas of southern Africa. They look superficially like a rabbit or a muskrat, but are nearly tail-less, have short rounded ears, and possess moist foot pads that provide a good grip on smooth rock faces. By outward appearances, these tiny hyraxes seem the most unlikely of candidates for being evolutionary cousins of massive elephants.

Yet, precisely such a relationship was proposed in 1884 by the famous paleontologist E. D. Cope. One key line of evidence was the soft padding on the hyraxes' feet, an unusual feature that is shared by elephants. However, any single feature of this sort could be misleading with regard to family ties, particularly when the organisms otherwise resemble each other so little. Are hyraxes truly related more closely to elephants than to other small mammals such as rodents and rabbits?

Incredibly, molecular data from many genes confirm what Cope originally suspected. In the mammalian evolutionary tree, hyraxes sit on a branch that is shared by elephants and their allies (including sea cows), rather than by rodents, rabbits, or any other candidate mammals such as warthogs, hippos, or antelopes. The elephant and the hyrax are not exactly kissing cousins—their ancestral lineages split more than 60 million years ago—but they do appear to be part of the same monophyletic (single-origin) group, or "clade," that originated in some distant shared ancestor.

A further surprise to molecular systematists was a discovery that the elephant-hyrax clade appears to be joined by several other African creatures formerly thought to be completely unrelated to one another. These include golden moles, aardvarks, and, as it turns out, the appropriately named elephant shrews. The latter are the size of a mouse but have long mobile snouts reminiscent of an elephant's miniaturized trunk. Altogether, this newly recognized clade of African mammals has given rise to a compelling theory called "Afrotheria," which posits that about one-third of all taxonomic orders of placental mammals originated from a single common ancestor that inhabited the Dark Continent about 75 million years ago.

From similar kinds of genetic evidence, the broader evolutionary tree of placental mammals displays at least three other such deep branches: one that includes rodents, rabbits, and primates; a second that includes bats, hoofed animals, and most carnivores such as cats and dogs; and a third composed of armadillos, sloths, and their allies. All these groups now are suspected to have split from one another and begun their incredible adaptive radiations during the latter stages of the age of the dinosaurs, in the Cretaceous period, some 65 million to 150 million years ago. This time period also coincides with the initial breakup of the southern supercontinent (Gondwanaland) by Earth's crustal movements. By separating and isolating Africa from South America and other landmasses in the Southern Hemisphere, this continental breakup probably played a pivotal role in producing those early forks in the mammalian tree.

Descendants of the early Afrotherian mammals were to flourish on the isolated African landscape, branching and flowering into a fabulous diversity of current species, ranging from elephants to hyraxes, with many forms in between. As is frequently the case, systematists had to look well beyond superficial appearances to get to the evolutionary root of this matter.

Brilliant Butterflies

Species that are unpalatable to predators often evolve bright colors advertising their noxious taste (see "Fabulous, Fabled Frogs," in Part 8). Such warning, or aposematic, coloration serves to alert the eater to the distastefulness of its potential dinner, thereby averting an experience that would be highly unpleasant for both parties. The disagreeable flavor of aposematic prey usually comes from noxious compounds in the foods that they themselves ingest, or that their bodies synthesize biochemically.

Sometimes, several prey species that inhabit the same area are composed of such unsavory characters. Then, natural selection promotes the evolution of shared warning colorations by all these potential targets, thereby reinforcing a blatant message to predators: "Don't mess with any of us!" Such jointly repugnant species that mime one another in their warning advertisements are termed "Müllerian mimics," and their evolved mimicry patterns can be astonishingly refined. A famous case in point involves several species of *Heliconius* butterflies, common inhabitants of tropical regions throughout Central America and northern South America, where they delicately flit about on gossamer wings. All *Heliconius* species are distasteful to birds and other potential predators.

Two of these lovely species are *H. erato* and *H. melpomene.* Each consists of multiple geographical races (recognizably different forms) that differ dramatically in wing-coloration patterns. For example, one wing-color race in each species has small yellow dots and elongated streaks displayed against a black background, another models large red blotches on the fore wings, and another sports brilliant iridescent streaks on both the fore wings and hind wings. Most remarkable, however, is the fact that analogous wing-color races in the two species show striking similarity in geographic range. For example, the range inhabited by the red-blotch race in *H. erato* is matched closely by the area occupied by a red-blotch race in *H. melpomene,* and so on for approximately twenty other racial pairs in the two species. Thus, at any geographic locale, members of the two species mutually reinforce the antipredator message.

Over what time frame did mimetic wing-color patterns evolutionarily converge upon one another in these gentle fliers? To begin to address this question, scientists conducted genetic surveys on gene sequences in the various wing-color races of *H. erato*. Analyses of these molecular data revealed a basal phylogenetic split that proved to distinguish all populations living east versus west of the Andes Mountains, a finding that makes considerable geographic sense because the Andes must be a major barrier to gene exchange in this lowland species. The magnitude of the genetic divergence indicates that these two geographic forms separated about two million years ago.

In contrast, by the same yardstick of genetic divergence, different populations of *H. erato* within either of these two major geographical areas were nearly indistinguishable. Yet many of these populations belong to different races of *H. erato* as gauged by the butterflies' distinctive wing-color patterns. What all this indicates is that the radically different wing-color motifs in this species must have emerged quite rapidly and recently (probably within the last 200,000 years) during the evolutionary process. Evidently, evolution in these butterflies' dazzling colors has been quick, and at each geographic locale there has been a rapid convergence in wing-color patterns between *H. erato* and *H. melpomene*.

Thus, the molecular genetic data once again have provided a useful phylogenetic backdrop against which to interpret the evolutionary history of other kinds of organismal features (in this case, predator-deterring wing-color motifs). In a remarkably short evolutionary time, these beautiful butterflies have come up with defensive strategies that both literally and figuratively are brilliant.

Flocks of African Fishes

Lake Victoria, bordered by Kenya, Tanzania, and Uganda in east-central Africa, is the world's second largest freshwater lake. Occupying a "rift valley," opened in the earth by recent continental shifts, it forms part of the headwaters of the River Nile so ardently sought by David Livingstone and Henry Stanley in the 1800s. For biologists, Lake Victoria and other rift-valley lakes in the region are also famous for their so-called "species flocks" of fishes.

More than 300 species in the family Cichlidae (cichlids, for short) inhabit Lake Victoria. Most occur nowhere else in the world, and collectively they display a striking array of feeding adaptations. There are species that graze algae, crush snails, feed on plankton or organic debris, prey on insects, dine on fish, and rasp scales (for food) from other fishes' tails. Some even acquire food by taking advantage of "mouth-brooding" habits. In many cichlid species, adults carry their babies in their mouths for protection. Other cichlid species, known as paedophages, often engulf the snout of a mouth-brooding female and force her to jettison her brood, which the paedophage then consumes.

Such diverse behaviors help to explain why so many different cichlid species can reside jointly in Lake Victoria. Another important factor is that body-color patterns and courtship routines unique to each species inhibit hybridization between the different cichlid forms. However, such present-day differences in ecology and mating behavior beg the following questions. How old, in evolutionary terms, is this exuberant species flock, and how did the cichlids differentiate into so many species? To help address such issues, several research groups have surveyed Lake Victoria's cichlids using numerous molecular markers.

Most of the species proved to be extraordinarily close genetically, and phylogenetically distinct from cichlids in several other East African lakes. Thus, it was in recent evolutionary time (well within the last million years) that Lake Victoria's cichlids evolved into such an astonishing number of species and lifestyles. This genetic finding begs another question. Did the speciation events occur within the confines of the lake itself?

Under "allopatric speciation," which is how species normally arise, an ancestral population first becomes separated into two or more geographic units. In the absence of genetic exchange, the daughter populations diverge by natural selection, genetic drift, and other evolutionary forces, and eventually may evolve into distinct species. This is the undisputed mode of allopatric speciation for most animals. But if the cichlid species in Lake Victoria truly arose within the lake, this might be an unprecedented example of prolific "sympatric speciation," wherein populations diverge into separate species without the initial benefit of physical barriers to dispersal.

Geologically, Lake Victoria is only about one million years old, a date generally consistent with the genetically deduced maximum age of its species flock. However, the lake's configuration has fluctuated greatly over time, and its bed dried almost completely as recently as 15,000 years ago. Thus, the possibility remains that many of Lake Victoria's cichlid species arose via microgeographic speciation at times when the lake was subdivided into multiple ponds. In other words, perhaps the species flock arose via allopatric speciation events after all, but on an unusually small spatial scale.

This possibility is extremely difficult to eliminate for rift-valley lakes, but another type of African lake has provided additional clues. Volcanic crater lakes are tiny and ecologically monotonous, each conical crater being a simple bowl with steep walls. Thus, there always must have been a lack of physical barriers to fish movements within any such lake. Yet, some crater lakes house miniflocks of cichlid fishes too. For example, Lakes Barombi Mbo and Bermin in Cameroon, western Africa, cover just a few square kilometers each, but nonetheless harbor eleven and nine endemic cichlid species, respectively.

Genetic analyses of these crater-lake flocks (plus fish in surrounding rivers) indicate that the species within each volcano diversified recently from a common ancestor. Although this would seem to clinch the case for sympatric speciations, even here competing scenarios are hard to disprove entirely. Perhaps closely related species from separate rivers independently colonized the lakes but then went extinct in their former habitats. Whether sympatric or allopatric speciation was involved, it is safe to conclude that these volcanic miniflocks of fishes erupted very recently in evolutionary time.

The African lakes have been marvelous evolutionary playhouses. Sadly, their biological dramas are in jeopardy. The Nile perch, an introduced predator, now decimates native cichlids in the rift-valley lakes. Another concern is an increase in water turbidity due to deforestation, agricultural runoff, and urban pollution. Murky waters cause failures in the fishes' mate-choice behaviors, which are based primarily on visual cues. Increased hybridization may follow, thereby eroding previously evolved differences and in effect destroying much of the preexisting biodiversity in the lakes. Thus, tragically, human actions are imperiling both the cichlid species themselves and the ecological theaters that promoted these fabulous evolutionary plays.

Microbats and Megabats

Upon first inspection, bats (order Chiroptera) would seem to be a tight-knit evolutionary lot that unquestionably shared a common ancestor (that is, are monophyletic). After all, they are unique among all mammals in possessing the capacity of flapping flight. A bat's wings are really its hands, covered by double membranes of skin that stretch between the finger and hand bones and extend to the forearm, sides of the body, and the hind legs. The skillful use of these appendages, in conjunction with sonar navigation, gives these mammals gifted piloting skills when they depart from their roosts.

Nearly 1,000 living species of bats are known, but basically there are two kinds: the familiar microbats (small, nocturnal insect-eaters that fly about from dusk to dawn, avoiding obstacles and catching their prey via echolocation, even in total darkness); and the megabats (large fruit-eaters that fly by day and would be quite helpless if out much after sunset).

Microbats and megabats differ in other ways as well. For example, most microbats can lower their body temperatures and may hibernate for long periods, whereas megabats lack this physiological ability. Another difference involves detailed features of neuroanatomy. According to some interpretations, the neurological features of megabats are more primatelike than batlike, and this suspicion gave rise to a "diphyletic," or "flying-primate," hypothesis, which proposes that megabats really are the evolutionary relatives of primates (like us) rather than of microbats. If so, the suite of adaptations associated with powered flight would have evolved from nonflying ancestral mammals at least twice: once in the primate lineage and once in the microbat line.

Are bats monophyletic or diphyletic? Based on molecular analyses of several marker genes, megabats and microbats have proved to be related more closely to one another than to primates. This finding is consistent with the conventional wisdom that the general bat morphology and a capacity for powered flight arose only once in the evolutionary history of living mammals. It also suggests that the neuroanatomical likeness of megabats and primates is due to convergent evolution.

Thus, after gaining flight more than 50 million years ago, some ancestral bat species apparently began an adaptive radiation that eventuated in the mega- and microbats of today. Notably, the latter group was able to exploit nighttime aerial niches that largely had been neglected by its day-active counterparts, the avian aviators.

Anomalies and Paradoxes
in Sunflowers

Can the process of hybridization sometimes lead to the rapid creation of a new biological species? Why not? After all, when the distinctive and finely tuned genomes (sets of DNA) of two species are thrown together in a hybrid organism, who knows what may happen? If the hybrid survives and is not made sterile by the genetic ordeal, it could produce offspring (grandchildren of the hybridizing parents) with a wide range of "recombinant genotypes" never before seen, some of which might be reproductively incompatible with those of the two parental species. Through successive generations of descendants, as things sort themselves under known rules of heredity and the culling effects of natural selection, a neophyte species might emerge in short order from this heterogeneous genetic mix.

That, in a nutshell, is the controversial theory of hybrid speciation. Is there any solid evidence that new species occasionally do arise in this fashion? As already mentioned in "The Lizard That Dispensed with Sex," in Part 2, hybridization between species *is* the normal route by which new unisexual (all-female) species originate, but perhaps these are special cases. At issue here is whether hybridization sometimes gives rise to new sexual species as well.

Helianthus annuus and *H. petiolaris* are common, well-known sunflower species widely distributed across North America. By contrast, three other sunflowers, appropriately named *Helianthus paradoxus, H. anomalus,* and *H. neglectus,* were poorly known localized species in the western United States, usually confined to very dry locales such as sand dunes. From morphological and distributional evidence, some scientists suspected that these three species with the intriguing scientific names might have arisen from hybridization between *H. annuus* and *H. petiolaris*. However, a critical test of this idea awaited the application of molecular markers.

One expected genetic signature of a hybrid species is that it should display an amalgamated mix of genetic markers otherwise confined to its two presumed parental taxa. However, such observations alone would not prove hybrid origins because, oddly, the species in question instead might be ancestral to its candidate "parents." To decide between the competing hypotheses of hybrid origin and ancestral polymorphism, the phylogenetic arrangements of genetic traits must be determined in the relevant taxa.

Through painstaking analyses of more than 100 molecular markers, it was shown conclusively that *H. paradoxus* and *H. anomalus* really are of hybrid origin, whereas, paradoxically, the no-longer-neglected *H. neglectus* is not. These genetic studies provide some of the clearest documentation that new sexual species in nature indeed can arise via hybridization, but also that not all suspicious taxa arose via the hybrid route. Because many other plant species likewise have dubious modes of origination, similar kinds of genetic analyses should provide a clearer view about the broader prevalence of hybridization in the speciation process.

Snapping Shrimps Another way that molecular markers can inform evolutionary biology is in "simply" describing how many living species exist. Not all reproductively isolated species can be distinguished readily by morphological or behavioral features. Many are cryptic. Genetic markers are useful in describing new species, particularly when the molecules demonstrate the presence of two or more separate gene pools in the same geographic area (in sympatry) where formerly only one was suspected to exist. A case in point involves some tiny crustaceans.

Have you ever snorkeled or scuba-dived on a tropical reef? If so, you may have noticed incessant clicking noises. They were the drummings of undersea percussionists, a busy bunch of little crustaceans known as snapping shrimps.

These tiny territorial shrimps live in holes and crevices on the reef, sometimes deep inside the labyrinthine passageways of living sponges and corals. The popping sounds are produced by an enlarged claw that each shrimp brandishes like an oversized six-shooter, always cocked and at-the-ready to snap at passers-by or to stun intruders. Thousands of shrimp live densely packed on most reefs, so their collective claw snappings produce constant static, like some poorly received underwater radio.

Snapping shrimps have given science two major surprises. First, like many hymenopteran insects (see "Extreme Social Behavior and Gender Control," in Part 4) and the naked mole rat (see "The Naked Mole Rat," in Part 5), some species of snapping shrimp are colonial and highly social. In such species, more than 300 offspring, all from a single queen, can be crammed into a single large sponge, cooperatively defending the colony. Advanced sociality (eusociality) is rare in the animal world, so its detection in a marine shrimp was totally unexpected. Another surprise came from the recent discovery of several formerly unrecognized species.

Traditionally, systematists lumped many morphologically variable forms of snapping shrimp into a fairly small number of distinct species. For example, although displaying a variety of colors and body sizes, some of the most common snapping shrimps in the Caribbean were thought to belong to a single species, *Synalpheus rathbunae*. Molecular genetic studies have shown that this taxon, in truth, is a complex of at least three distinct gene pools. Further field observations showed that each of these valid biological species is associated with a different sponge host. Thus, taxonomists previously had underestimated the true number of shrimp species in this complex by at least several-fold.

Similar genetic appraisals of numerous marine invertebrates, including various groups of sponges, corals, mollusks, and bristleworms, suggest that many cryptic species lurk in the sea. Will naming and describing these taxa be merely an academic exercise, a mundane cataloguing operation? Far from it. Reliable descriptions of species provide a critical foundation for any scientific inquiry into nature's workings. Furthermore, proper recognition of the true number of living species can provide a firmer basis for conservation efforts.

Valid taxonomies also can have direct economic implications. For example, in attempting to identify and extract valuable organic compounds from sea creatures, pharmaceutical companies have a vested interest in proper catalogues of ocean life. Correct taxonomies and phylogenetic alignments of living organisms in the sea (and elsewhere) are like biological road maps that enable a more intelligent exploration of nature's otherwise bewildering wilderness.

10

Wildlife Forensics and Conservation

F orensic medicine" is defined (in the *Cambridge English Dictionary*) as "the application of medical knowledge to the elucidation of doubtful questions in the court of justice." Molecular genetic markers, in particular, can provide critical evidence in many kinds of courtroom situations, as, for example, in paternity suits or murder trials. By analogy, "wildlife forensics" might be defined as the application of genetic evidence to doubtful questions in the court of nature. Almost all uses of molecular markers in ecological, behavioral, or evolutionary studies at least touch on issues of forensic identification. Here, "molecular wildlife forensics" will mean the express use of genetic data to identify particular animal or plant products where such diagnosis otherwise is difficult or impossible. Such genetic diagnoses are often relevant to conservation efforts.

This part will illustrate how genetic markers have been used in diverse forensic applications: assigning dissociated body parts to the creatures from which they have come, designating to which species ambiguous larval forms belong, distinguishing hybrid from nonhybrid organisms, diagnosing morphologically cryptic species, and recognizing endangered species. Especially when illegal trade in threatened species or their biotic products is involved, some of these forensic applications can even help to settle doubtful issues in national and international courts of human law.

The Plight of the Whales
Centuries of systematic hunting, exacerbated by the invention of steam-powered vessels and the exploding harpoon in the twentieth century, have driven many species of whales and their allies (order Cetacea) to perilously low levels. Tens of millions of these intelligent marine mammals formerly roamed the world's oceans, but many species today number only in the few thousands of breeding individuals. Among the magnificent species currently listed as endangered or threatened are several baleen species, including the humpback and blue whale, who, despite their great size, filter tiny plankton for food; and the famous sperm whale, the toothed species that challenged Captain Ahab yet normally attacks and eats squid.

With many cetacean species rapidly approaching extinction, the International Whaling Commission, a multinational regulatory body, voted in 1982 to impose an indefinite moratorium on commercial whale hunting. Only a few exemptions were granted, for example, when specimens were needed for scientific purposes, or as objects of subsistence hunting by aboriginal cultures with established whale-harvesting traditions. Yet, nearly twenty years after the moratorium went into effect, whale meat ("kujira" in Japan, "gorae" in Korea) continues to show up in abundance in retail outlets, particularly in eastern Asia and Scandinavia. How can this be?

One possibility is that the "whale meat" comes from smaller cetacean species such as porpoises and dolphins, which still can be harvested legally (even if unethically) in some countries. Or perhaps the retail material is not from cetaceans after all but, rather, represents compressed fishmeal or other seafood products that have been mislabeled to appear more exotic or pleasing to consumers. Another possibility is that the whale meat came from large cetaceans harvested legally under an International Whaling Commission exemption, or from long-frozen specimens that had been killed decades earlier when the practice remained legal. Of course, the substantive concern is that much of the meat might come from ongoing illegal harvests by unscrupulous hunters.

What, really, are kujira and gorae? To find out, geneticists purchased samples of "whale meat" from numerous retail outlets in Asia. These products had been advertised as dried salted strips of whale meat, or as slices of raw whale meat, skin, and blubber. Taste tests alone could not determine the source of the material, but molecular forensic analyses proved that these samples came from a variety of cetacean species. About half of the retail samples were from populations of whales or dolphins that plausibly had been harvested under legal permits. However, other retail samples included meat from endangered humpback and fin whales, and from an ostensibly protected population of minke whales in the North Atlantic.

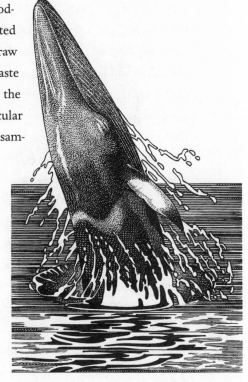

These findings helped document that the illegal killing of whales continues to this day. Thanks to these new genetic techniques, it now should be possible to monitor the blackmarket activities of the whale plunderers and thereby better assess the biological effects of their unconscionable actions. Thus, if our generation is not to witness the further demise of cetaceans, law enforcement must become a high priority, as should additional treaties and aggressive international agreements to protect most of the remaining populations of whales, porpoises, and dolphins. If there is an encouraging note on which to end this discouraging tale, it is that more people than ever are aware of and concerned about the plight of some of the grandest and most intelligent creatures ever to have inhabited the planet.

Pinniped Penises
The global trade in wildlife products (both legal and illegal) is huge, with the estimated value of the "industry" ranging from $8 billion to $20 billion per year. The inexcusable overhunting associated with wildlife commerce has seriously depleted or already caused the extinction of numerous species. Depending on the type of organism, the sought-after commodities may be specialty foods, pelts, shells, ivory tusks, feathers or other ornaments, or many other animal goods. However, perhaps no highly sought product in the commercial wildlife trade is more surprising than an animal's sexual organs.

Seals, sea lions, and walruses (suborder Pinnipedia) constitute the majority of the world's marine mammals other than whales and porpoises. Although distantly related to dogs, cats, and other carnivores (flesh-eaters), pinnipeds have front and hind flippers that make them powerful swimmers, and thick layers of fat that insulate them against the cold. Pinnipeds may migrate far out to sea, but they typically live along coastlines, hauling themselves ashore to rest and give birth. It is when on land that pinnipeds are most vulnerable to human hunters. Several of the species are sought for meat. Walruses have been hunted for their valuable tusks. Others, such as the fur seal, are prized for their luxurious

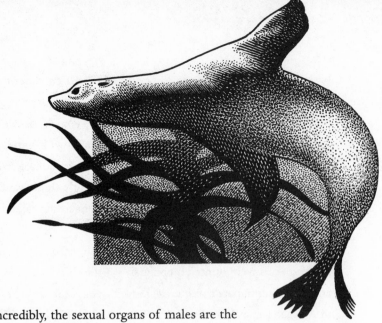

coats. Incredibly, the sexual organs of males are the basis for yet another hunting tradition.

Particularly in Southeast Asia, a thriving trade in pinniped penises (along with the associated baculum bones and testes) is supported by a folklore that these sexual organs are powerful aphrodisiacs. These private parts of seals are sought so avidly that a single set of gonads, sold as dried tissue, powder, or as whole organs soaked in wine, can command $50 or more in traditional Chinese medicine shops. Sex organs have become the most valuable product derived from these animals, and many thousands of seals are butchered each year for sale in international markets. For example, in 1993 alone, Norway shipped approximately 8,000 harp seal penises to Hong Kong, thereby supplying about 50 percent of that island's marketplace demand.

In retail outlets, it is usually difficult or impossible to identify, by visual inspection alone, the particular species from which gonadal tissue has been taken, and this greatly complicates any contemplated attempts to monitor and potentially regulate commercial trade in these animal products. Which species are affected most severely by the international market for gonadal aphrodisiacs? Scientists recently extracted DNA sequences from each of about thirty gonads purchased from traditional Chinese medicine shops in Shanghai, Hong Kong, and Bangkok in Asia, and from San Francisco, Vancouver, Toronto, and Calgary in North America. The molecular information, when compared against existing genetic databases for carnivorous mammals, permitted the assignment of each gonadal sample to a particular mammalian species or taxonomic group.

Although advertised as pinniped penises, the gonads for sale in these medicine shops proved to have originated from a wide variety of animals, including wild dogs, domestic cattle, and water buffalo, in addition to harp seals, hooded seals, and fur seals. Evidently, there was only partial truth-in-advertising with regard to the material being sold to the consumer.

In earlier centuries (and sometimes continuing today), various pinniped species were hunted mercilessly for their flesh or pelts. For example, indiscriminate commercial harvests nearly caused the extinction of Alaskan fur seals in the Bering Sea, and elephant seals in the North Pacific were nearly eradicated when overhunting in the late 1800s caused their numbers to plummet to fewer than thirty animals (all clinging to existence on one remote island west of Baja California). Fortunately, those species somehow survived the onslaught, and at least for the time being are stable or recovering.

Today, many pinniped species are either protected legally or managed for sustainable harvest. However, even as efforts to ensure the survival of seals and walruses have grown, so too have the conservation challenges. In particular, the despicable international market in sexual organs must be opposed and halted. To help combat this trade, relevant consumers must be educated to the biological fallacy of animal gonads as aphrodisiacs.

A Tasty Turtle Many people find turtle flesh delectable, and this sometimes has led to excessive hunting. For example, indiscriminate harvests of sea turtles over the last four centuries have helped to drive some of these species to their current status as threatened or endangered. More than 200,000 cases of canned turtle soup were processed in Key West, Florida, during the 1880s alone, and as recently as 1957 more than a million pounds of live sea turtles were imported into the United States for food. Serious population declines eventually prompted federal and international restrictions, and legitimate trade in marine turtle products finally ended in the 1970s.

Snapping turtles, with their long tails and short tempers, are among the largest of turtles in the freshwater realm. In particular, the alligator snapper *(Macroclemys temminckii)* of the central and southern United States reaches two feet or more in length, and old individuals can weigh in excess of 200 pounds. Alligator snappers often lie patiently at the bottom of rivers or lakes, jaws held wide open. Inside each animal's mouth is a curious pink protrusion that looks and wriggles like a small worm. This is the creature's fishing lure, and an unsuspecting fish will rue the day that it falls for the turtle's subterfuge.

Unlike marine turtles, freshwater turtle species in the United States have little protection through harvesting restrictions. Sizable domestic markets persist, to the point where the numbers of several species, including the alligator snapper, have declined severely. Accordingly, this species recently gained legal protection from commercial trapping in every state within its range, except Louisiana.

What fraction of the "turtle meat" sold in markets comes from the alligator snapping turtle? Biologists recently purchased and analyzed thirty-six putative turtle-meat products (frozen, cooked, and canned) from retail outlets in Louisiana and Florida. Upon close genetic inspection, only one of these

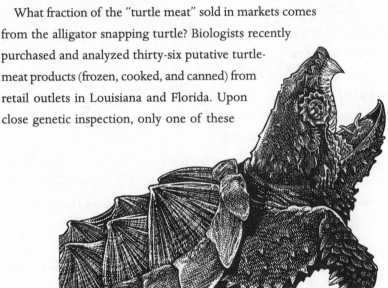

purchased samples proved to be of flesh from an alligator snapping turtle. Most of the rest had come from smaller, unprotected species such as the common snapping turtle and the Florida softshell. Interestingly, eight of the tested products (22 percent) that had been labeled as turtle meat were not from turtles at all, but instead were from American alligators.

Thus, the commercial marketplace for turtle meat exemplifies what can happen in unsustainable wildlife harvests. The largest, most abundant, or most prized species (such as green turtles and alligator snapping turtles in the current case) are exploited first. When their numbers decline sharply, less treasured species (such as the Florida softshell) are substituted, and so on in a downward-spiraling cycle. If left unregulated, this inevitable chain of commercial events serves the longer-term interests of no constituency—not the hunter, not the consumer, and certainly not conservation biology.

"Caviar Emptor" Sturgeons (family Acipenseridae) are long-lived fishes capable of reaching ten feet in length and typically achieving sexual maturity at ten to twenty years of age. These primitive-looking beasts sometimes are considered "living fossils," because similar forms were common in Jurassic seas, when dinosaurs ruled the planet nearly 200 million years ago. About twenty-five sturgeon species are alive today, but most of them are severely depleted in number and now are listed as endangered or threatened.

The fate of the Atlantic sturgeon, *Acipenser sturio*, illustrates the extreme plight of many sturgeon populations. Until the beginning of the twentieth century, this species was abundant in shallow coastal regions of the eastern Atlantic—from Iceland to the Mediterranean, and particularly in the North and Baltic Seas. It had been used for food since ancient Roman times, and records dating as far back as the sixteenth century indicate abundant catches in the Rhine estuary and other parts of western Europe. However, the harvest declined catastrophically in the early 1900s, and by the 1970s only an occasional specimen could be found on spawning runs in the Rhine or other European rivers. By the late 1980s, only about 500 individuals remained, nearly all in one small population in the Black Sea.

Sturgeons are the primary producers of esteemed "black caviar," to educated palates the most appetizing of appetizers and the true epitome of fish roe. Like fine wines, fish eggs are cherished for their delightful tastes, and three of the most prized varieties are beluga caviar (from the beluga sturgeon, *Huso huso*), osetra caviar *(Acipenser gueldenstaedtii)*, and sevruga caviar *(A. stellatus)*. These three sturgeon species seldom reproduce on their own any more, but they are bred artificially. The young, released into nature to grow, are harvested as adults years later. Sturgeon eggs can sell for $50 or more per ounce, and the high market value is a primary factor in the global decline of sturgeon populations because it promotes poaching and illegal harvests (in addition to the legal catch).

Which species of sturgeon provide the black caviars sold in retail outlets? To find out, investigators purchased nearly one hundred lots of caviar (in tin cans or glass jars), primarily from gourmet shops in the New York City area. Most had been labeled as beluga, osetra, or sevruga caviar and sold at high prices accordingly. By profiling genetic markers from the eggs, biologists showed that all lots indeed were bona fide sturgeon roe, but 23 percent had been mislabeled as to species. In most of these latter cases, roe from an endangered species of sturgeon had been substituted for eggs purportedly taken legally from the three major commercial species.

Thus, a vicious cycle is in place. The severe decline of sturgeons at human hands caused the short supply of roe that maintains the high prices for black caviar. In turn, the extraordinary commercial price of sturgeon eggs promotes the poaching of other sturgeon species, including those at immediate risk of extinction.

In response to this conservation crisis, an education campaign has been launched to help save the sturgeons. Its catchy slogan "caviar emptor" (a play on words of *caveat emptor,* "let the buyer beware") is meant to alert consumers to the serious repercussions of caviar demand. Will this or other conservation measures alleviate fishing pressures on sturgeon? Let us hope so, because if not, we may lose forever some very special creatures whose ancestors once frolicked with the dinosaurs.

An Endangered Bird in the Belly of a Snake

What is an "endangered species"? Under the Endangered Species Act of 1973, it is defined to include any severely depleted "subspecies" or any "distinct population segment" of any type of wildlife or plant that is in clear danger of extinction. In the United States, each such entity receives legal protection when its name is included in an official catalogue of endangered and threatened taxonomic groups, or taxa. This list is revised and updated continually. In concept (but not always in practice), creatures determined to be at severe risk are added to the list, and any population deemed to have recovered sufficiently is a candidate for removal from the compilation.

Nowadays, molecular genetic techniques are employed routinely to help identify endangered species. These new kinds of data are highly relevant to taxonomic decisions and, hence, to conservation efforts for jeopardized animals and plants. Sometimes, a genetically distinctive taxon is identified that warrants inclusion on the endangered species list. Other times, a taxon listed as endangered proves not to be highly distinct in genetic composition from nonendangered ones, in which case the formal listing may be called into question. Both kinds of outcomes are illustrated by genetic work on seaside sparrows *(Ammodramus maritimus).*

This brown-streaked bird inhabits salt marshes from southern Texas to New England. Geographic populations differ slightly from one another in plumage features, and this has led to taxonomic controversies. In particular, two native Floridian populations with noticeably different feather hues—the dusky seaside sparrow *(A. m. nigrescens)* on Cape Canaveral and the Cape Sable seaside sparrow *(A. m. mirabilis)* in the Everglades region—were listed as officially endangered. Dusky seaside sparrows have a darker plumage than most other seaside sparrows, and the feathers of the Cape Sable sparrow differ by having a slightly greenish tinge. How divergent in overall genetic makeup are these two sparrow populations from one another, and from other seaside sparrows elsewhere along the coast (which are not listed as endangered)?

Because of protective federal regulations concerning endangered species, geneticists were denied permits to kill any specimens for molecular analysis. In the case of the dusky seaside sparrow, genetic analyses had to await the natural death of the last living specimen, an elderly individual who passed away in a Florida zoo after attempts to establish a captive breeding colony failed. (Thus, despite being of academic interest, the genetic findings for the dusky sparrow were posthumous and moot—the "species" already had become extinct.) In the case of the Cape Sable

sparrow, the specimen available for genetic analysis was retrieved, as partially digested remains, from the belly of a large cottonmouth snake who had killed and eaten the unfortunate victim in an Everglades' marsh.

In any event, when finally assayed, both the dusky and the Cape Sable seaside sparrows proved to be extremely close phylogenetic relatives of other seaside sparrows along Atlantic shores but only distant cousins to sparrow populations along the

Gulf of Mexico. From this evolutionary-genetic perspective, the traditional taxonomy for the seaside sparrow complex was probably in error because it failed to reflect this deep genealogical split. The traditional taxonomy was also in error because it had emphasized population distinctions that were far more shallow or evolutionarily recent.

The genetic evidence suggests instead that formal taxonomic names (probably at the subspecies level) should be given to the two regional genetic clans of salt-marsh sparrow—those that generally inhabit Atlantic versus Gulf coastal regions. These distinctive genetic tribes first may have been separated by geographic barriers during the ice ages of the past two million years. Neither of these two genetic forms is in immediate threat of extinction today, but these are the principal evolutionary units within the seaside sparrow complex that should merit continued attention from conservation biologists.

The broader point is that taxonomic names can be more than just convenient handles to attach to organisms. They also can convey evolutionary-genetic information that is germane to biodiversity assessments, management plans, and conservation efforts.

The Ridley Riddles

From the fossil record, marine turtles first appeared about 150 million years ago. Seven or eight species are alive today, all officially listed as threatened or endangered. In the last three centuries, human activities have reduced most marine turtle populations by more than 90 percent. People have harvested turtle eggs excessively and have slaughtered countless numbers of adults for food, shell ornaments, or as "by-catch" in fishing industries directed largely at other sea creatures. For example, despite the recent requirement that "turtle excluder devices" be used by the shrimping industry in the southeastern United States, many adult turtles are killed each year when they drown in the shrimp nets pulled behind trawl boats. Other turtles die when caught on baited hooks deployed in longline fisheries for tuna or other oceanic fishes. Yet another serious concern for marine turtles is the loss of suitable nesting habitat due to beach alteration and shorefront development.

The Kemp's ridley *(Lepidochelys kempii)* is the most endangered of all marine turtles, in large part because it nests almost exclusively at a single site (in Tamaulipas, Mexico). However, a close relative of the Kemp's, the olive ridley *(L. olivacea)*, is the most abundant of the living sea turtle species, with a nearly worldwide distribution in tropical and subtropical waters. The two recognized ridley species are similar morphologically, and their lopsided geographic distributions seem to make little sense. Thus, some biologists doubted that the Kemp's and olive ridleys really were distinct species. If true, this would have important ramifications both for taxonomy and for conservation plans within the ridley complex.

However, in recent molecular analyses, these two recognized taxa proved to be quite different genetically—far more so than is typical for "conspecific," or same-species, populations of most other marine turtles. Thus, by genetic evidence, the recognition of two ridley species now appears fully justified. The following scenario, although speculative, may account for the original evolutionary separation of these two species, and also for their current geographic distributions.

About three million years ago, when the Isthmus of Panama first rose above the sea, an ancestral ridley species became subdivided into a proto-Kemp's form in the Atlantic Ocean and a proto-olive form in the Indo-Pacific. Before that time, an open-ocean channel existed between Central and South America that would have permitted the easy movement of tropical turtles between the Atlantic and Pacific Oceans. The newly arisen land bridge (that is now Panama) would have isolated these marine populations for the first time and thus could explain the relatively deep genetic separation exhibited today between the Kemp's and olive ridleys. Then, much later, olive ridleys recolonized the Atlantic Ocean from the Indo-Pacific (probably via dispersal around South Africa), thus accounting for the current-day presence of olive ridley populations in both major ocean basins.

For many years, the United States and Mexico have cooperated on an ambitious plan to help save the highly endangered Kemp's ridley turtle. Thousands of hatchlings have been collected from the Tamaulipas nesting site for transport and release on Padre Island in south Texas, the intent being to establish a second nesting colony that will reduce the risk of species extinction by any site-specific catastrophe such as a hurricane or an influx of predators. In other words, biologists hope that Kemp's ridleys soon will be laying their eggs in more than one rookery basket. The recent genetic findings have bolstered justification for this noble conservation effort by showing that Kemp's ridley turtles truly are distinct in evolutionary terms from their closest living relatives.

Ridleys also are unique among marine turtles in that they have a spectacular mass-nesting behavior, wherein, once a year during daylight hours, legions of females come ashore all at once to lay eggs. When this behavioral phenomenon was discovered in the Kemp's ridley in 1947, more than 40,000 females blanketed the Tamaulipas beach on a single day. In surveys conducted about four decades later, only a few hundred turtles returned annually to that same shore. Clearly, the Kemp's ridley is in serious straits, so let us hope for resounding success in ongoing recovery efforts for this very special ancient mariner.

Lakes, Swamps, and Gene Pools

Lake Chatuge is a man-made reservoir in the upper reaches of the Tennessee River drainage, along the border between Georgia and North Carolina. Long renowned for excellent bass fishing, the reservoir is monitored carefully and its fish populations are censused annually by professional fishery biologists.

Lake Chatuge originally was populated by native smallmouth bass *(Micropterus dolomieui)*, but in the 1980s something odd took place. In annual surveys, biologists noticed that smallmouth bass were in sharp decline. In their stead were captured increasing numbers of spotted bass *(M. punctulatus)*, a related species not native to the Tennessee River. Furthermore, some of the fish were hard to classify, being generally intermediate in appearance between pure smallmouth and pure spotted bass.

Nobody is certain where the spotted bass came from, but one popular fisherman's tale is that this species was introduced into Lake Chatuge by members of a bass club returning from a successful fishing trip to Lake Lanier in the Chattahoochee River drainage, where spotted bass are common. In any event, by the early 1990s, fish generally resembling *M. punctulatus* constituted more than 90 percent of the total bass catch from Lake Chatuge.

Exactly what were these newcomer bass in Lake Chatuge, and what had happened to the formerly thriving population of smallmouths? Genetic markers have yielded the answers. It turns out that nearly all bass now in Lake Chatuge are either genetically pure spotted bass, or hybrids between spotteds and smallmouths. Among the hybrid fishes examined, a few were first-generation progeny, but many more were "backcross" specimens (those that arose from crosses between hybrid fish and genetically pure spotted bass).

Following the unauthorized introduction of spotted bass into Lake Chatuge, the conversion of the reservoir from a smallmouth bass fishery to a spotted bass/hybrid fishery was amazingly fast. Apparently, the original spotted bass and their hybrid and nonhybrid progeny were better suited than smallmouth bass to the ecological conditions in the lake, and within just a few years they simply took over.

What became of the native smallmouth bass? This is a clear example of local genetic extinction mediated by the introduction of an exotic species. In this case, however, the immigrant was a close relative of the native species, and hybridized with it extensively. Today, more than 95 percent of the smallmouth bass genes that remain in Lake Chatuge are present not in the few pure smallmouths that remain, but rather in fish of hybrid ancestry. Biologists say that the smallmouth genes have been "genetically assimilated" into the spotted bass population.

Hybridization between spotted and smallmouth bass in Lake Chatuge also had an interesting sporting consequence. Bass can reach a large size, fight ferociously against light tackle, and are sought avidly by fishermen. Lake Chatuge is a spotted bass paradise, and has been home to several state-record specimens. The latest of these trophies, caught in 1994, weighed nearly nine pounds. However, at the certification station, visual inspection of that fish raised some skeptical eyebrows, and frozen tissues were sent to a genetics laboratory. Analyses of this animal's proteins and DNA revealed that the fish was actually a first-generation hybrid between a spotted bass female and a smallmouth bass male. Because the fish was not a pure spotted bass, the fisherman's once-in-a-lifetime catch was denied official certification as a state record.

The Fish Whose
Babies Get All Mixed Up
Rockfish (genus *Sebastes*) are a morphologically and ecologically diverse group of more than 100 species mostly confined to the cold waters of the North Pacific Ocean, particularly along the American west coast. These slow-growing fish have unusually long lifespans of several decades. Various species inhabit settings ranging from shallow kelp beds and rocky outcrops to silty depths below 1,500 feet.

Some rockfish species are active swimmers, with streamlined bodies of dull pink and brown. These forage off the bottom in large schools. Others are sedentary and territorial, with deep heavy bodies and large heads armed with elevated ridges and poisonous spines. Some of these species display stunning colors. For example, the tiger rockfish is barred in bright red and black, the treefish is striped in yellow and black and has pink lips, the quillback is dressed in burnt orange and black, and the China rockfish flaunts a lovely band of gold along its otherwise dark body.

Rockfishes have internal fertilization, and females later give birth to live young. Before settling, these tiny larval fish swim about in the plankton for long periods, sometimes a year or more. Baby rockfishes are nearly impossible to identify as to species by appearance alone, but collectively the mixed batches of larvae can account for as much as one-third of all fish specimens captured when a plankton net is towed through ocean waters.

To better monitor reproductive success and recruitment patterns, fishery biologists would love to be able to identify rockfish babies as to

species. But how might this be done, since all of these larvae look pretty much alike?

Despite the large number of rockfish species in the North Pacific, most proved to be readily distinguishable from one another by their genes. These molecular findings indicate that marine rockfishes have accumulated genetic differences over a relatively long period of evolutionary time. This also means, at least in principle, that the look-alike babies in the plankton can be identified as to species by molecular assays. Someday this genetic approach may be scaled up to help biologists determine the proportions of various rockfish species that are represented in a given horde of planktonic larvae.

Such genetic information in turn might assist in developing fishing regulations. Rockfishes collectively have supported some of the largest commercial and sport fisheries on North America's west coast. However, few of these fisheries will be sustainable for long under traditional rates of harvest, and further conservation measures are needed urgently. One small but important step will be in learning which rockfish species are reproducing most successfully.

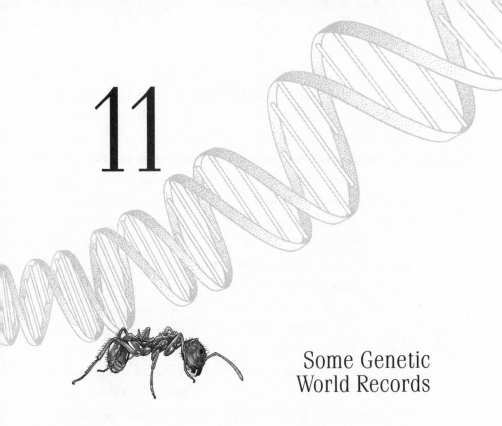

11

Some Genetic World Records

Previous sections explained how genetic natural historians have gained many important insights into nature's operations—findings that are of academic interest in their own right, and in some cases also have practical ramifications for such areas as biological monitoring and conservation efforts.

The case studies in this part make no special claim for broader scientific relevance or enlightened social policy. They simply report some particularly curious genetic findings from a miscellany of plant and animal species. However, if a "Genetic Book of World Records" were ever to be produced, these creatures and the stories they tell might well qualify for inclusion.

The 100-Ton Mushroom

What is the world's oldest and largest living creature? One serious contender lurks in the mixed-hardwood forests of the Upper Peninsula of Michigan. The beast lives near Crystal Falls (close to where I often played, unsuspecting, as a child). No, it's not one of the region's many black bears or white pine trees, and it's not Paul Bunyan's fictitious giant blue ox, Babe. It's not a wolf, but it does creep stealthily along the forest floor (albeit slowly, at a rate of about one foot per year). It hides mostly underground, but rears itself up occasionally, especially during the warm summer months. It wears a cap of reddish brown, its body is a yellowish wash, and it has gills that are a pinkish buff. Well camouflaged in its woodland lair, it's a fungus—the honey mushroom, *Armillaria bulbosa*.

Most mushrooms reproduce sexually when their familiar aboveground forms make millions of wind-dispersed spores, but many species also can spread vegetatively by cordlike aggregations of "hyphae" that look like extremely long shoe laces. These weave through the soil and across its surface, occasionally sending up new fruiting bodies that we notice as distinct mushroom heads. Any above-ground mushrooms that have arisen from the same network of interconnected hyphal threads must ultimately trace back to a single fertilized egg, or zygote. Thus, all the cells, hyphae, and mushroom heads that make up such an aggregation would be extended parts of the same genetic body. They would all be a part of the same "individual."

A remarkable discovery, resting on DNA evidence from multiple genes, is that all honey mushrooms gathered across a forty-acre patch near Crystal Falls proved to be absolutely identical to one another genetically. Thus, they are an extended part of one-and-the-same genetic individual that must have spread across that huge area by long-term vegetative growth. Based on observed rates of hyphal extension and mushroom biomass per unit area today, calculations indicate that this individual started life as a fertilized egg about 1,500 years ago and that it now weighs at least 100 tons in aggregate.

Phylogenetically, fungi are related more closely to animals than to plants. The largest animal ever to have inhabited the planet, the blue whale, can weigh as much as 150 tons, somewhat larger than our fungal friend in Michigan. In the plant world, the oldest known living tree is a gnarly, weather-beaten bristlecone pine, affectionately named Methuselah. It sprouted in the White Mountains of eastern California about 4,700 years ago. One of the largest individual tree specimens thus far documented is a giant sequoia, also in California, weighing in at about 1,000 tons. Upper Michigan's honey mushroom isn't quite as ancient or grandiose as these solitary plant giants, but nonetheless, this special fungus living among us should rank high on any list of Earth's oldest, biggest, and most venerable creatures. And, even more elderly and gargantuan fungal specimens no doubt await discovery.

Copepods: Nature's Most Abundant Animal?
What is the most abundant species on Earth? It's probably a virus or some other microbe such as a bacterium, whose population sizes can be astronomical. Among multicellular creatures that are visible (if sometimes only barely) to the naked eye, various species of marine copepods are particularly abundant. Take, for example, the tiny crustacean *Calanus finmarchicus*, a free-swimming relative of crabs, and one of about 5,000 described copepod species. About thirty individuals laid end to end would barely stretch an inch, but what this organism lacks in size it makes up for in prodigious numbers. This copepod lives in the open-water oceanic realm, forming part of a mega-pool of planktonic species (minuscule drifting organisms) that provide food for many fishes and whales. Indeed, an adult right whale consumes more than 2,500 pounds of plankton per day.

Based on counts of *C. finmarchicus* in small samples of seawater, and considering the total estimated volume of ocean inhabited by this species, conservative calculations suggest that today's population in the North Atlantic consists of about 1,000,000,000,000,000 specimens (that's one million multiplied by one million, then multiplied again by one thousand).

Has this copepod's population always been so huge? In principle and sometimes in practice, molecular data permit estimates of prehistorical population sizes. This is because the expected evolutionary depth of genealogical lineages within a species is related to how many near and distant ancestors led to the descendants alive today: all else being equal, the more ancestors, the deeper the species' family tree. In other words, all genes that survived to the present trace their lineages to a finite number of ancestors, and the larger that number, the longer ago some of the genetic lines split from one another. All this "coalescent theory" can get pretty mathematical, but the basic idea is simple. In the case of *C. finmarchicus*, empirical data from DNA lineages, interpreted in the context of coalescent theory, indicates that the species in effect has been composed on average of only about 100,000 breeding individuals per generation across recent evolutionary time. This is vastly less (by 10-billion-fold) than the current number of living specimens. Quite likely the species has experienced severe reductions in population size at least periodically, perhaps due to disease outbreaks or changing climatic regimes, for example. For these or other reasons, the species cannot have been mega-abundant continuously across evolutionary time, because the present-day lineages then would be much older than has proved to be the case.

There are many possible gauges to evolutionary success. We humans tend to think of intelligence as the epitome of evolutionary adaptations, but it also may promote our early extinction. Horseshoe crabs (see

"Horseshoe Crabs," in Part I, on page 9) might argue that longevity in geological time makes them the most successful creatures on Earth. Bristlecone pine trees (see "The 100-Ton Mushroom," above) might claim that species whose individuals live longest should be declared winners of the evolutionary game. Copepods might assert that trophies for the champions of evolution should be reserved for species with the largest numbers of living individuals.

However, the molecular genetic data on copepods (as well as most other currently abundant animals similarly assayed) indicate that population sizes in these species have not always been nearly so large. Thus, by the standard of numbers of individuals, the evolutionary success enjoyed by any species might well be fleeting.

Mating Champions of the Insect World Among

the insects, what is the world's champion female in regard to the number of successful mating partners? It's hard to say, because females (as well as males) of many insect species mate multiple times during their brief but hectic lives. For example, on her nuptial flight a female wasp may copulate with as many as ten different males. In the blue milkweed beetle, a female mates up to sixty times, although not always with different partners.

Presumably, females receive and store sperm from each mating event, but acquiring sperm and then using that sperm to fertilize eggs are two different things. Thus, the number of copulations, even when with different males, may not be an accurate register of the true number of different sires who eventually contribute to a female's pool of offspring. So, what

species holds the record for most male partners with whom a female has mated to produce progeny?

According to a recent review of the scientific literature, the current record holder appears to be the giant honeybee *(Apis dorsata)* of Malaysia. From genetic evidence on paternity (by DNA fingerprinting techniques), an average queen mated successfully with about thirty males, because that many sires are required to account for the number of different paternally derived genotypes observed among her progeny. Remarkably, one giant honeybee female was documented to have had conjugal alliances with fifty-three drones who fathered her offspring.

In general, honeybees are extremely active on their nuptial flights. Copulations occur on the wing when a virgin princess is approached by a squadron of males (a "drone comet") at an altitude of about sixty feet. First, one male clasps her royal highness in a dorso-ventral position and everts his phallus, ejaculating sperm from a special bulb that figuratively explodes into her vagina. For his efforts, the drone is paralyzed in the process and falls backward, forfeiting a portion of his sexual organ. Tumbling to the ground, he dies. Undeterred by this grotesque outcome, another suitor clasps and mates with the princess, and so on in a spectacular aerial orgy. Loaded with sperm, the newly anointed queen then flies off from the sexual battlefield, littered with corpses below, to establish her own colony. The large number of successful matings in the giant honeybee is all the more remarkable because the average total duration of a drone comet, at dusk, is a mere thirteen minutes in this species.

Lonesome George: The World's Loneliest Beast?

When Charles Darwin arrived in the Galápagos Islands in 1835 as a naturalist aboard HMS *Beagle*, land tortoises were common, and each island had its own distinctive form, or race. Why were there slightly different types of tortoise for each island, and why did populations in the "Galápagos" (a Spanish word for "turtle") differ even more so from other tortoise species on mainland South America? As Darwin pondered the creative forces that might have produced such evident variety, he was led gradually toward revolutionary views of nature's operations. The land tortoises on various islands of the Galápagos differed visibly from one another, albeit subtly, because of the cumulative effects of natural evolutionary forces, acting through time, on these semi-isolated populations. Furthermore, these differences must have arisen sometime after colonizer tortoises from the South American continent first arrived in the Galápagos archipelago.

Recent phylogenetic analyses, based on DNA sequences, have revealed that the Chaco tortoise *(Geochelone chilensis)* of west-central South America is the continental species related most closely to *G. nigra*, the Galápagos species. Furthermore, the genetic data indicate that these two species last shared a common ancestor about six to twelve million years ago. Presumably, the Galápagos Islands (or their now-drowned precursors) were colonized in these ancient times by continental waifs that rafted to the islands, pushed by a northwest-flowing oceanic current. Once established, various populations on the Galápagos isles diverged from one another and from their mainland cousins.

Even before Darwin arrived on the islands, the Galápagos tortoises were in decline from decades of human abuse. Because these sluggish, long-lived tortoises could be stored live in a ship's hold for months on end, the half-ton giants were an ideal source of fresh meat for early buccaneers and whalers. More than 200,000 animals were slaughtered, beginning in the seventeenth century. At least as serious was the depredation due to introduced pests such as rats and dogs,

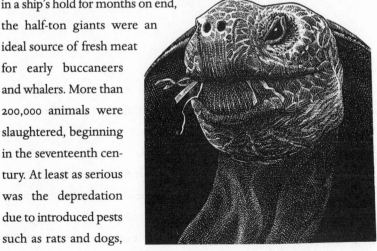

which destroyed the tortoises' eggs. Only in recent times have humans begun to value these endangered animals, not only for their historical place in the development of evolutionary thought, but also for their sheer size, beauty, and venerability.

All of which brings us to an old bachelor known as Lonesome George, arguably the longest lonely heart on Earth. Many decades old, George is a 650-pound giant Galápagos land tortoise, lonesome because he is the last surviving member of the Pinta Island race *(abingdoni)* that was thought to have been eradicated early in the 1900s. To make George happy (truthfully, to repopulate Pinta Island with land tortoises most like those that were native to the island), biologists would like to find George a suitable mate so that together they might produce offspring. Among the surviving races of *G. nigra* on other islands in the Galápagos chain, which one might provide George with the most desirable bride?

Past efforts to interest the old bachelor in mating have failed—over the years, George has shown no enthusiasm for females placed in his captive enclosure. However, recent genetic assays have shown that these earlier matrimonial candidates were from island races quite different from that to which George belongs. Actually, George's closest evolutionary relatives inhabit the islands of Española and San Cristóbal at the southeastern end of the Galápagos chain. Armed with this new phylogenetic information, biologists now can introduce George to candidate brides perhaps more to the liking of this old lonely heart.

Life's Earliest Farmers

About 10,000 years ago, humans invented "agriculture"—a botanical domestication process wherein the seeds, cuttings, or tubers of desirable wild plants purposefully are sowed at chosen prepared sites, and the resulting foods or fibers later reaped in organized harvests. Over the millennia, selective breeding and the refinement of farming practices increased the reliability and abundance of food supplies far beyond what had been offered by earlier hunting-and-gathering traditions. These agricultural achievements in turn enabled the formation of dense urban assemblages (the first cities), the rise of human societies with stratified class structures and specialized work forces (including professional militia to defend and acquire land), and the organization of societies into elaborate political states and countries. In other words, farming permitted and prompted "civilization."

Given the tremendous reign that humans now hold over the planet, this invention of agriculture a scant ten millennia ago must rank among the most influential events in the history of life. However, human beings were not nature's first successful farmers. Another group of animals already had that title. These are the attine ants (tribe Attini), who have farmed domesticated mushrooms for more than 50 million years.

More than two hundred species of these highly social New World ants cultivate fungal gardens upon which they completely rely for food. The process starts when a queen ant, about to depart on her nuptial flight, packs into her mouthparts a small wad of mycelia (filamentous fungal threads) taken from the garden of the colony in which she was born. After mating in the air, she dives to the ground, casts off her wings, digs a nest in the soil, plants her own fungal garden, and begins to lay eggs. The first worker ants emerge in about six weeks and begin to tend the fungal garden upon which the colony's survival will depend.

The ants' agricultural practices are elaborate. To prepare the soil and nurture symbiotic fungi, these tiny social creatures continually harvest great quantities of leaves that they chew into a nourishing, pulpy mush. They regularly take fungal cuttings and knead the filaments into the soil of newly prepared plots within the nest. The ants also weed their gar-

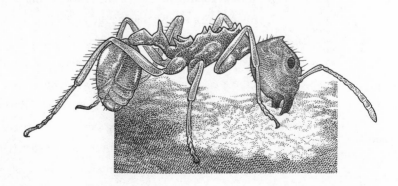

dens by plucking out nondomesticated fungal species and by secreting antibiotic "herbicides" to control molds. They religiously fertilize the garden soils with their fecal fluids and later harvest the fruits of their labors to feed themselves, the queen, and the colony's developing larvae.

The evolutionary history of ant farming has been unraveled recently from analyses of various DNA sequences. Based on these molecular genetic data, phylogenetic trees were reconstructed for numerous species of attine ants as well as their symbiotic fungi. By comparing these two phylogenetic trees, the following discoveries were made: (a) the ants collectively have succeeded at farming several different types of fungi; (b) the fungal domestication process is ongoing in some ant species; (c) the ants are capable of switching to novel fungal crops; and (d) a given type of fungal food crop occasionally is transferred between unrelated colonies of ants.

The many operational parallels between ant farming and human agriculture are remarkable indeed. If ants rather than humans were writing textbooks on the history of civilization, they would rightly emphasize that an important transition from hunter-gathering to organized agriculture had been initiated by some of their forebears more than 50 million years ago.

How Low Do Root Tips Go?

Molecular genetic markers sometimes come to service in the most unlikely of situations. Have you ever wondered how deep into the earth a plant's roots extend? Actually, as concerns many trees especially, that's quite a challenging question. Normally, you can't just pull one up, like a carrot, to check directly, and even if you could do so with heavy logging equipment, many of the finer roots that reach the greatest depths would be ripped off in the process. Such difficulties are exacerbated for trees growing in heavily compacted soils, or those whose roots meander among subterranean rocks.

The depths of tree roots are not of interest only to trivia buffs. Botanists and soil scientists have a keen interest in the topic because plant-root depths register and influence the movements of water and nutrients in the soil and also are associated with patterns of plant productivity. Indeed, unseen below-ground competition among plants for moisture and nutrients may be at least as intense as their more obvious above-ground jostling for space and light.

To determine how deep some roots may go, a genetic study was conducted on various trees on the Edwards Plateau of central Texas. Woodlands there are dominated by several species of oaks, junipers, elms, sugarberries, and mesquite. At varying depths underground, the region is infiltrated by myriad limestone caves into which small tree roots protrude through the ceiling from above. By visual inspection alone, however, it is impossible to know exactly which root belongs to which tree. The key to the genetic analysis was to match tiny tree roots snipped from inside caves to a particular tree above ground from which they had grown. Molecular markers did the trick.

Through such genetic analyses, it was discovered that at least six different tree species in the area grow roots downward for more than fifteen feet. The champion miner was the evergreen oak *(Quercus fusiformis)*, whose roots sometimes probed cracks and crevices to a depth of seventy-five feet or more. This meant that some of these trees were "taller" underground than they were above ground.

The surface soils of the Edwards Plateau are shallow, hold water poorly, and are nutrient sparse. Thus, there can be fertile rewards for trees that drill deeply for water and nutrition (including, perhaps, the nitrogen-rich

guano deposits inside bat caves). Although there is little evidence that trees can "sense" underground resources at a distance and grow purposefully toward them, adventurous roots nonetheless can sometimes reap great benefits.

A Fish Returned from the Dead

What animal species holds the record for seeming resurrection from the most ancient of evolutionary graves? It's the coelacanth, *Latimeria chalumnae*. Coelacanth fishes arose about 350 million years ago and are represented in the fossil record until about 65 million years ago, when they suddenly went extinct along with the dinosaurs. Well, that's what biologists used to think.

Fossil coelacanths were large prehistoric fish, up to six feet long, with thick bony scales, heavy-set lower jaw, and a huge symmetrical tail. They also possessed fleshy, bone-supported fins that resembled incipient legs. Indeed, scientists suspect that coelacanths constitute a side-branch of an evolutionary line of fishes that gave rise, hundreds of millions of years ago, to the earliest backboned animals with limbs. As these proto-amphibians crawled from primordial seas, they embarked on a grand adventure that was to eventuate eons later in the evolutionary emergence of reptiles, birds, and mammals.

In one of the most exciting biological discoveries of the twentieth century, in 1938 a living coelacanth was trawled from deep waters off the mouth of the Chalumna River in eastern South Africa. The ship's crew was unaware of the significance of their find, but at dockside a biologist, Marjorie Courtenay-Latimer, was astounded by the specimen and brought its existence to the attention of science. A kind

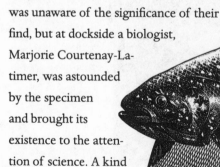

of fish that was thought to have been extinct for 65 million years had risen miraculously from the dead. The species later was named in honor of its finder and the locale from which the coelacanth had been taken.

Fourteen years later, a second specimen was caught near the Comoro Islands (between Madagascar and Mozambique), and then several more fish were turned up in this same general region of the western Indian Ocean. In 1998 another coelacanth population was discovered in Indonesian waters, more than 6,000 miles away. Clearly, viable populations still exist in the Indian Ocean, and underwater photographers actually have captured these piscine beauties on film.

Thanks to the unexpected availability of tissue samples from living coelacanths, geneticists were able to address a long-standing phylogenetic puzzle that otherwise might have remained unsolved forever: Where exactly does the ancient coelacanth lineage fit in the evolutionary tree of backboned animals? Extensive DNA-sequence assays have confirmed that coelacanths (and other lobe-finned species such as lungfishes) actually are related somewhat more closely to amphibians, reptiles, birds, and mammals than to most other fishes. These genetic findings do not mean that the historical relationship between coelacanths and mammals is intimate. However, it does mean that these resurrected creatures give us a modern glimpse of what some of our distant ancestors may have looked like just before they first crawled out from the primordial seas.

The World's Most Shocking Marriage?

Prince Charles of England surprised the world when he married a commoner, Lady Diana. But this wasn't the first or the most shocking of such marriages between royalty and the lower classes. Genetic assays recently have uncovered another such European wedding whose overall effect shook the world even more.

The exceptional wines from the Burgundy and Champagne regions of France have been cherished for centuries. Pinot Noir, Gamay Noir, and Chardonnay are among the genetic varieties (cultivars) of the grape, *Vitis vinifera*, that has given the world some of its finest red and white table wines as well as many delightful champagnes. These and other French grape varieties are among the true aristocrats of the viticultural realm. Or are they?

Wine grapes under cultivation are propagated vegetatively, so all vines within a particular cultivar are clonal, genetically identical to one another (barring any mutations that may have arisen since that strain's origin). All domestic grapes stem ultimately from wild grape ancestors, but various popular cultivars prized today are thought to have arisen by several mechanisms: separate domestications from native grape vines, natural crosses between wild vines and cultivars, and spontaneous crosses between pre-existing cultivars. Genetic analyses indicate, for example, that Cabernet Sauvignon, arguably the most highly regarded red-wine grape in the world, arose from an unforced cross between two other noble grape varieties under culture—Cabernet Franc and Sauvignon Blanc.

What are the historical origins of the splendid Chardonnay and Gamay cultivars? These two blue-blooded varieties of grape both display genotypes indicative of hybrid origins. Furthermore, the genetic markers indicate that the hybridization events were between the cultivars Pinot and Gouais Blanc. The exact number of separate crosses involved is uncertain, but the general outline of the family tree (or vine) for these grapes is now quite clear.

One of the parents in this cross, Pinot, is not surprising, because this cultivar itself is a true aristocrat with a long and famous history in France perhaps dating to times of the Roman conquest. However, this cultivated prince of grapes crossed with Gouais Blanc, a variety of grape considered so pedestrian that periodically it was banned by legal decree in parts of the country. Indeed, its name derives from "gou," an old French word of derision.

Gouais Blanc may have been introduced to France from eastern Europe in the third century, perhaps as a gift to the Gauls from the Roman emperor Probus, an avid viticulturist. During the Middle Ages, Gouais Blanc was one of the most common grapes in the vineyards of northeastern France, but it was confined to mediocre locations, with the superlative sites reserved for Pinot and other more noble varieties. Who could have guessed that some of the world's finest wines stem from hybridization events involving Gouais Blanc?

The Planet's Most Common Vertebrate?

The copepods (see "Copepods: Nature's Most Abundant Animal?" above) are one of the most abundant animals on Earth today, but among all creatures with a backbone (fishes, amphibians, reptiles, birds, and mammals), what species holds the record? Incredibly, despite our large body size, humans are somewhere on the short list. More than six billion people currently inhabit the planet, and we are increasing by an astonishing net tally of 260,000 individuals per day (10,800 per hour, 180 per minute, or 3 per second). No wonder we're overrunning the planet, using a grotesquely disproportionate share of its resources, and crowding many other species out of existence.

But if we really want to know which vertebrate species is the most abundant numerically, we probably should look to much smaller crea-

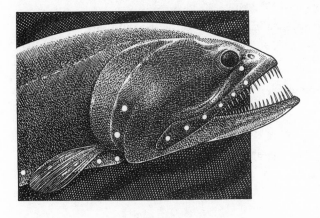

tures in the open oceans. After all, the seas cover about 70 percent of the planet's surface and provide more than 90 percent of its inhabitable volume. Particularly vast are the mesopelagic and bathypelagic realms, those unfathomably large briny depths below 100 fathoms (600 feet). Situated between the thin veneer of the illuminated ocean surfaces, and the even thinner sheet of bottom habitat on the ocean floor, these dark mid-water zones constitute the greatest bulk of potentially livable space on the planet.

So, who among the backboned creatures lives in this vast mid-water realm? Well, lots of different kinds of fish certainly do. For example, bristlemouths and lightfishes turn up consistently in deepwater trawls. As their names imply, these fish have, respectively, finely toothed jaws and rows of luminescent spots along the body that light the inky darkness like portholes. One species of lightfish in particular, the inch-long *Cyclothone alba*, can be collected in huge numbers throughout the Atlantic, Pacific, and Indian Oceans. Accordingly, at least one author (Nelson) has nominated the cyclothone fishes as Earth's most plenteous type of vertebrate organism.

Geneticists recently examined populations of this cyclothone fish from around the world. With respect to DNA sequences, these samples showed larger genetic differences than normally are present within a single species. Furthermore, each oceanic basin or subbasin appeared to have its own distinctive genetic form. These molecular genetic findings strongly suggest that at least five different species formerly had been masquerading under the name *Cyclothone alba*.

Given the logistic difficulties of catching and censusing fish from the deep sea, it's anyone's guess as to exactly how many individual lightfish are present in the world's oceanic depths. For the sake of argument, let's say that the number for *Cyclothone alba* is about a kazillion. From genetic evidence, however, we now suspect that this taxon consists of at least five separate species. Thus, the number of each of those species may be more like a kazillion divided by five, but it's still a pretty big number.

12

Fossil DNA

Genetic inferences about the recent or distant past normally are based on comparisons of DNA or protein molecules from living species. But thanks to an incredible technological breakthrough in the mid-1980s, biologists now can rouse small pieces of DNA literally from the dead. Even a few short molecules of DNA exhumed intact in the fossil remains of some long-extinct organism sometimes can be copied or amplified to assayable levels using a powerful new laboratory technique known as the "polymerase chain reaction," or PCR.

The PCR is a biochemical chain reaction in which even a few strands of intact DNA prime the production of more and more copies of themselves. The procedure is not unlike conventional photocopying except that the facsimiles in this case are duplicated DNA molecules identical in structure to the originals. After enough new copies of a particular DNA sequence have been synthesized artificially, in a test tube, they can be assayed just as if the sequences had been purified in high numbers from a rich living tissue source such as blood or liver.

The PCR in effect can bring low-quantity or comatose DNA molecules back to life. However, when applied to amplify the tiny smidgens of DNA that at best are found in exceptionally well preserved fossil materials, the PCR procedure can be tricky. Nonetheless, in special circumstances, ancient DNA sequences can be retrieved successfully from the body parts of creatures that died even tens of thousands of years ago.

The evolutionary tales in this concluding section will illustrate how geneticists are reading the genetic messages in well-preserved fossil DNA to unearth novel information about the evolutionary affinities, dietary habits, and other surprising attributes of some rather magnificent species that long ago went extinct.

Ancient Asphalt Jungles Downtown Los Angeles might

seem an unlikely site for one of the world's richest fossil deposits. Yet here, among the city's modern asphalt jungles, lie some ancient asphalt beds of nature's own design. These are the famous Rancho La Brea tar pits, from which have been unearthed a fabulous variety of fossilized remains of creatures that inhabited southern California 10,000 to 40,000 years ago.

These tar pits formed when crude oil seeped through fissures in the earth's surface. Gradually, the light fractions of the oil evaporated, leaving behind a sticky goo or asphalt concoction that over the millennia has been a deadly mire for unsuspecting animals. Nearly 2 million fossil specimens representing 660 species (many long since extinct) have been recovered from the La Brea mortuary. Paleontologists calculate that the number of biological remains in this mass graveyard is consistent with an average of about one major entombment episode every ten years throughout the 30,000-year duration of the La Brea tar pits.

Many of the animals cached in this natural cemetery are giant predators or scavengers. Among the birds are numerous condors, eagles, and extinct storklike creatures known as teratorns. Even more impressive are the large mammals, including giant ground sloths, dire wolves, tapirs, camels, peccaries, mammoths, mastodons, and American lions. These and many other fossils in the heart of downtown Los Angeles provide a poignant reminder of how dramatically humans have altered the North American landscape and its flora and fauna over the last 10,000 years.

The cast of mammalian characters in La Brea is filled with impressive species, but perhaps none more so than the extinct saber-toothed cat *(Smilodon fatalis)*. Roughly the size of an African lion, this muscular feline's most notable feature was its long pair of knifelike teeth used to stab, slash, and slice open hapless prey.

Fossil remains of saber-toothed cats recently have contributed to our genetic knowledge of this extinct species. After entrapped animals died in the asphaltic syrup, their soft parts quickly degraded. However, the bones, despite being infiltrated by natural petroleum, apparently protected cellular DNA. Using special extraction techniques, scientists were able to retrieve, PCR-amplify, and then assay short DNA sequences from the bones of giant cats preserved at the La Brea site.

Gene sequences from several saber-toothed cats were compared with those of nine living species in the cat family Felidae. Phylogenetic analyses revealed that the extinct saber-tooths were embedded within the "pantherine" evolutionary lineage that also includes living lions, tigers, leopards, jaguars, and cheetahs, all species that are thought to have evolved from a common ancestor within approximately the past 10 million years. This genetic scrutiny of the tarry remains of saber-toothed cats provides a remarkable demonstration of how modern laboratory methods can uncover some intriguing new findings from old fossils.

Genetic Visions of Mammoths

Many people alive today have seen an artist's rendition of a woolly mammoth. But just a few thousand years ago our own ancestors knew these elephantine beasts as living creatures. That's because woolly mammoths, and their kin the mastodons, were common inhabitants of Pleistocene landscapes, and our recent forebears hunted them or scavenged their carcasses for meat. Indeed, human hunters from Asia, who colonized the New World little more than 12,000 years ago by crossing the Bering land bridge to Alaska, may have been a major factor in the subsequent extinction of mammoths, mastodons, and many other large herbivores in prehistoric North America.

Woolly mammoths *(Mammuthus primigenius)* generally resembled living elephants but had a more steeply sloping back, proportionately shorter hind legs, and huge semicircular tusks that curved back nearly to the base of their trunk. Mammoths' massive, heat-conserving bodies also had thick coats of long, shaggy fur, enabling the behemoths to withstand subarctic climates during the Ice Age. This also meant that when an animal died, its remains sometimes were flash-frozen and preserved in nature's northern iceboxes.

Today, people again are on the hunt for woolly mammoths. This time, scientists are searching for frozen carcasses, encased in ice or permafrost, that might yield DNA suitable for detailed examination of the evolutionary relationships of these extinct creatures to living Asian and African elephants. Several such fossil remains have been uncovered, and they are remarkable finds indeed. Unlike most animal fossils, which are mere imprints in sedimentary rock or mineralized bones, some of the preserved materials from mammoths include actual skin, fur, and soft tissues.

As you must have surmised by now, molecular genetic assays have been conducted on some of this finely maintained material. Analyses

of several DNA sequences retrieved from frozen mammoths (and mastodons) have documented that these extinct animals were an extended part of the tightly knit lineage that includes living Asian and African elephants. Thus, they also proved to be cousins of dugongs, manatees, hyraxes (see "Evolutionary Trees and Elephants' Trunks," in Part 9), and other more distant relatives of the Elephantidae.

Some geneticists have more grandiose visions for the frozen tissues. Some woolly mammoth carcasses might include cells with exceptionally well preserved genomes. If so, the notion is that these DNA blueprints for mammoths might be implanted into a properly prepared egg from an elephant, and the egg then returned to the uterus of a living female. If all went well (a big "if"), months later the pregnant mother elephant would give live birth to a baby woolly mammoth. Until a few years ago, such fantasies would have seemed like utter science fiction. However, Dolly the lamb was engineered genetically in nearly this identical fashion (albeit using genetic material from the cells of a living ewe). If scientists indeed do proceed with attempts to regenerate a live mammoth, whether successfully or not, at least no one can accuse them of thinking small.

Facts on Bruin Evolution

Another extinct mega-mammal from which DNA has been retrieved is the cave bear *(Ursus spelaeus)*. This fearsome creature, even larger and more powerful than living brown bears *(U. arctos)*, ambled the Pleistocene landscapes of western and central Europe until as recently as 20,000 years ago. This giant's grinding molar teeth, with broad chewing surfaces, indicate that despite the animal's intimidating size and demeanor, the cave bear was primarily a vegetarian.

Where does the extinct cave bear fit in relation to the family tree of living brown bears? Using the PCR method, DNA sequences reportedly were amplified from cave-bear bones preserved in French grottoes for more than 25,000 years. The genetic data accumulated from these fossil DNAs then were compared with DNA sequences from living brown bears sampled across Europe.

Living populations of brown bears in Europe proved to be subdivided into two major genealogical (historical) units, one confined to central and western portions of the continent, and the other, perhaps of Asian origin, found farther to the east. The molecular genetic data also indicate that these two branches in the family tree of Old World brown bears separated about one million years ago. By these same molecular yardsticks, the cave-bear lineage likewise appears to have separated at roughly this same point in evolutionary time.

Another interesting fact about bruin phylogeny has come to light from similar molecular analyses of brown bears living in the New World. There, in western North America, two highly distinct lineages also exist, each with its own restricted geographical distribution. But a big surprise is that one of the genetic lineages in brown bears also includes another species, the polar bear *(U. maritimus)*.

Polar bears, of course, are snow-white bears that roam the frozen ice fields of the Arctic. In outward appearance and in ecology and behavior, they would seem to have little in common with the brown bears of warmer climes to the south. Nonetheless, the genetic results indicate that Arctic polar bears have extremely close genealogical ties to brown bears of southeastern Alaska. Indeed, their genetic lineages may have separated a scant 200,000 years ago.

At least two possibilities could explain this outcome. Perhaps polar bears are an ancient species but recently hybridized with Alaskan brown bears and thereby received some brown-bear genes. In captivity, polar bears and brown bears occasionally do produce viable and fertile hybrid offspring. Alternatively, the ancestors of polar bears may have broken family ties with brown bears quite recently in evolution and subsequently diverged rapidly in general appearance and behavior as they adapted to the extreme conditions of the Arctic. More research is needed to determine which of these two scenarios actually accounts for the genetic ties between these two distinctive species of bruins.

The Diets of Sloths

Two groups of sloths survive today in the rainforests of Central and South America: the three-toed (genus *Bradypus*) and two-toed *(Choloepus)* forms. Appropriately named, sloths are lethargic animals usually observed hanging upside down, their long toes secured around a tree limb. The living sloths are distant cousins of armadillos and anteaters, all members of the mammalian order Edentata that began to diversify about 80 million years ago.

However, the sluggish sloths of today's rainforests had many closer relatives as recently as 10,000 years ago. Numerous sloth species, placed into forty genera and three taxonomic families, were common beasts during the Pleistocene epoch before going extinct (some undoubtedly at the hands of human hunters) near the end of the last ice age. Many of these species were tree dwellers, but others lived on the ground, and some of the latter were real behemoths, reaching the size of large bears and elephants. As testimony to their dominance, they often left behind splendid skeletal remains.

From fossil bones and teeth of two extinct species of ground sloth *(Mylodon darwinii* and *Nothrotheriops shastensis)*, researchers extracted genetic material and compared the DNA sequences with those of living Edentata. Results confirmed a close relationship between the living and extinct sloth species, and their more distant alliance with armadillos and anteaters. But another fossilized heirloom passed down from the ground sloths—their feces—proved to yield even more scientific information.

In the southwestern United States and elsewhere, specimens of ancient dung ("coprolites") have been found in dry caves and rock shelters that sloths apparently used as latrines. Remarkably, DNA molecules can sometimes be extracted from these stools, and their sequences read to reveal what the animals had been eating. Sloths clearly were herbivorous, so the relevant molecules to be examined in the fossilized excrement were those (cpDNAs) normally present in the chloroplasts of plants.

By comparing the "fossil cpDNA" molecules with those from living plant species, the sloths' menu items were deduced. The genetic analyses revealed that these sloths ate salads of capers and mustards, lilies, grasses, saltbushes, cucumber-flavored herbs, mints, mallows, and grapes. Who would have dreamed that such dietary information could be extracted genetically from excremental material passed on by creatures who themselves passed on more than 10,000 years ago.

"Coproscopy" is the fancy word that adds a touch of class to the scientific analysis of ancient manure. But by any name, this discipline's subject matter can yield some interesting genetic discoveries.

The Demise of the Flightless Moa

The "ratites" are an ancient group of flightless, terrestrial birds often reaching a huge size. Among the living, these include the rheas of South America, the ostrich of Africa, the cassowaries and emu of the Australian region, and three species of kiwis in New Zealand.

The distribution of ratites across the world's southern realms suggests that this avian group may have diversified from a common ancestral species following the ancient breakup of the supercontinent Gondwanaland. This great landmass of the Southern Hemisphere existed throughout a good portion of the age of the dinosaurs but then began to fractionate, the pieces gradually drifting apart to become today's Africa, South America, India, Australia, New Zealand, and Antarctica. As the proto-continents separated, so too presumably did the ancestors of flightless birds. Recent molecular findings generally are consistent with this hypothesis, because the living ratite species show exceptionally large genetic differences that must have accumulated over a long period of evolutionary time.

Of special interest to evolutionary biologists are the ratites confined to New Zealand. Apart from kiwis, these also included (until recently) about fifteen to twenty species of wingless moas, some of which reached three times the weight of an ostrich.

Tragically, the Polynesians who colonized New Zealand a thousand years ago quickly drove all these plant-eating giants to extinction. About the only lasting good that came from this unconscionable slaughter is that ancient Maori campgrounds are prime sites today for recovering bones, eggshells, feathers, and bits of skin and dried flesh from the deceased moas.

Geneticists have isolated DNA from these fossils and compared their sequences with those from kiwis and other living ratites. The results indicate that moas and kiwis are not one another's closest kin. Instead, the kiwis sit on a branch of the ratite family tree that includes the emu, ostrich, and cassowaries as distant cousins. These findings imply that ratite birds colonized New Zealand at least twice: once in an evolutionary line that led to moas, and a second time in a genetic lineage that produced the kiwis.

Moas and kiwis aren't the only flightless birds to have evolved in the relatively safe New Zealand environment, which until human arrival was completely devoid of mammalian predators. Also native to the islands were grounded ducks, rails, wattled crows, parrots, and even a tiny wren. Alas, all these flightless birds (as well as many flying species) soon were exterminated, many by the Polynesians and more by later European colonists. Some of the extinctions resulted from overhunting and habitat destruction, whereas others occurred via the carnage from introduced dogs, cats, rats, and other human associates.

Having destroyed what they found on these isolated islands, people in the last two centuries have introduced untold numbers of European plants and animals in their stead. So massive has been this biotic turnover that if you visit New Zealand today, you will see mostly European flora and fauna rather than native New Zealand species. Such is the astonishing pace at which humans have undone what continental drift and evolution took a hundred million years to accomplish.

The Quagga Quandary

The domesticated versions of the horse *(Equus caballus)* are descendants of a species closely resembling the highly endangered Mongolian wild horse *(E. przewalskii)* that still gallops in perilously small herds across the plains of central Asia. They and their evolutionary cousin, the zebra, together with the wild asses (for example, *E. africanus*) and the half-ass (yes, there really is such a species— *E. hemionus* of the Middle East), make up the taxonomic family Equidae.

Actually, there are several living species of zebra, each with a different pattern of barring on the body. For example, the black-and-white stripes on the Burchell's zebra extend onto the belly and are relatively broad, whereas those on the Grevy's zebra are truncated on the flanks and narrower, like some dizzying test pattern on a television screen. A zebra's stripes tend to break up its body outline, presumably making these animals more difficult for lions and other predators to see.

All of which brings us to the quagga *(E. quagga)*, a lovely creature native to the desert areas of Africa. In outward appearance, the quagga looks part horse, part zebra. It has the intense barrings of a zebra on its head and neck, but these fade on the flanks and blend into a plain creamy-brown on the legs, and on the hindquarters which set off a bushy white tail. Tragically, the quagga is no longer extant. In the 1870s the species was driven to extinction in the wild, mercilessly hunted down because it competed with domestic goats and sheep for scant vegetation in its arid environment. The last quagga died in captivity in an Amsterdam zoo in 1883.

Although the quagga has been extinct for more than a century, some pieces of its DNA survived to present times in the dried skin of a stuffed museum specimen.

Molecular comparisons of these DNA fragments with those from living species of Equidae have shown that the quagga was a zebra rather than a horse, and that it was related most closely to Burchell's zebra.

This genetic finding has prompted the Quagga Project. Initiated by the South Africa National Museum, this project's incredible goal is to regenerate quagga-resembling animals by selective breeding. Burchell's zebras alive today show considerable variation in the intensity and distribution of stripes, and some have noticeable quagga-like features and coloration. By selectively mating such animals, it is hoped that a quasi-quagga population can be reconstituted and released back into the wild. Indeed, in 1999, the first three captive-bred "quaggas" from this ongoing project were liberated in South Africa's Karoo National Park.

Neanderthals and Us
Humans *(Homo sapiens sapiens)* are products of nature too, a very late arrival on the evolutionary stage. Extensive genetic assays have revealed a great deal about our genealogical origins and affinities. For example, they demonstrate conclusively that our closest living relatives are the great apes, in particular the common and pygmy chimpanzees *(Pan troglodytes* and *Pan paniscus)*. Molecular data indicate that the proto-human and proto-chimp lineages parted ways about 5 million to 7 million years ago, shortly after the stem lineage itself had separated from another hereditary pathway that eventually led to living gorillas *(Gorilla gorilla)*.

Another discovery abundantly supported by molecular evidence is the extremely close genetic relationship among human skin-color races. Indeed, our genes give little indication that the concept of "race" in humans has validity. Instead, under our skins, we all are surprisingly alike despite our superficially varied exteriors. This fundamental unity underlying human diversity is a consequence of the recency with which humans populated the planet from a common ancestral source. Genetic data suggest that there was a migrational exodus from the African continent as recently as 200,000 years ago. This modernity of human an-

cestry accounts for the genetic closeness of all peoples around the world.

For the six-million-year time interval between the split of the human-chimp lineages and the global peopling of the planet 200 millennia ago, fossils have provided most of the information on human origins. These reveal a rough progression of various named hominid forms (for example, *Ardipithecus, Australopithecus, Paranthropus, Homo habilis,* and *Homo erectus*) until full-fledged *Homo sapiens* emerged approximately 400,000 years before the present.

However, one group of recent fossil hominids remains an enigma: the Neanderthals *(Homo sapiens neanderthalensis)*. Anthropologists distinguish Neanderthals by their heavy brow ridges, less pronounced chins, and slightly larger brain sizes (on average) than modern humans. Neanderthals first appeared in Europe and western Asia several hundred thousand years ago and persisted until about 30,000 years ago, thus overlapping greatly in time with *Homo sapiens sapiens* (us).

Did the Neanderthals interbreed with our immediate ancestors, and do we still carry some Neanderthal genes? An exciting scientific breakthrough occurred recently when researchers extracted a piece of intact mtDNA from the well-preserved arm bone of a Neanderthal who lived more than 30,000 years ago in Europe. Direct comparisons revealed that this individual's mtDNA sequence falls well outside the range of genetic variation observed in modern humans. The magnitude of this genetic difference suggests that the Neanderthal's maternal lineage split from "ours" about 600,000 years ago.

The current molecular evidence also suggests that Neanderthals may have disappeared without contributing extensively to the genetic heritage of modern humans. However, these data must be interpreted cautiously because the mtDNA fragments recovered were short, and many more fossil specimens and genes must be examined to assess the full scope of genetic variation in the Neanderthals. Might it be the case that we still carry a great many genes from our former Neanderthal brethren? Although these creatures are now extinct, many of their genes may well live on, in us.

Epilogue

I hope you have enjoyed this introductory tour through nature's enchanting boutique and the brief narratives on some of its evolutionary exotica. In truth, however, nature is far more than just a curio shop of miscellaneous biological knickknacks. Left undisturbed, each creature in the wild is part of a functional "eco-*system*," one patch in a complex quilt of biological interactions, one filament in a biotic warp and weft that has been woven over the eons by natural selection and other evolutionary forces. In a very real sense, an organism removed from its natural surroundings is no more complete than an organ surgically removed from a donor's body.

Biologists have named nearly two million living species of animals and plants, and the total number in existence may be tenfold higher. How many of these distinctive forms of life can be saved from extinction in the coming decades? More important, how many can be preserved as integrated, functional components of natural ecosystems? These are open questions, among the most compelling biological issues of the twenty-first century. The planet currently is in the midst of a mass extinction forced by human overpopulation. Through man's unprecedented impacts on the environment, the fabric of life is fraying rapidly. How many more threads can be lost before the tattered cloth unravels? Each extinction of a species is a tragedy that impoverishes our world functionally, aesthetically, and informationally. Each death of a species at human hands is a moral indictment against our kind.

It was not my conscious intent to focus on threatened and endangered species in the case histories presented in this book. Nonetheless, in looking back, at least thirty-eight of the ninety-two essays (41 percent) deal primarily or exclusively with species of special conservation concern or that already went extinct. A bias in this compilation may exist because many of the essays describe large attractive animals, like pandas, or cute little creatures, like honeycreeper birds, and such well-known and loveable species are represented disproportionately on official lists of threatened taxa. The vast majority of truly endangered species, however, are poorly known (and often undescribed) forms that don't show up in the official tallies. Many of these inhabit biotic-rich environments such as coral reefs and tropical rainforests that are being destroyed even before their species can be discovered and named. Regardless of how the tabulations are conducted, the fact remains that the biotic world is under extreme duress.

In describing some of nature's intimate genetic secrets, I have had a broader purpose—to engender a greater appreciation of Earth's biological treasures. With increased awareness can come only heightened admiration, enhanced affection, and a deeper commitment to preserve what remains. We must learn to cherish the exuberance of life not only for its functional worth but also for its emotional and aesthetic value and for the intellectual gifts it offers. Every creature on Earth is, literally, one of our genetic kin, an evolutionary cousin on some distant branch of our extended family tree. By comprehending and savoring the remarkable diversity yet unity of life, we gain a wiser understanding not only of nature's operations but of ourselves.

Appendix

Molecular Genetic Techniques

Contrary to Hollywood images, most geneticists don't peer endlessly through microscopes in their biological detective work. The features of DNA and protein molecules that provide genetic markers are far too small for direct visual inspection, even under the highest of magnifications. Instead, researchers deduce the properties of genes indirectly, through the use of molecular and biochemical methods. Many such laboratory techniques exist, and most of them involve four principal steps: (1) extract DNA (or proteins) from blood or tissue samples, (2) physically separate different classes of these molecules, (3) make visible the separated forms, and (4) score the molecular differences in genetic terms.

What follows are brief descriptions of various laboratory methods that were employed singly or jointly in all the case studies highlighted in this book. Interested readers can find further methodological details and procedures of genetic data analysis in my book *Molecular Markers, Natural History and Evolution* (New York: Chapman and Hall, 1994) and in *Molecular Systematics*, 2nd edition, edited by D. Hillis, C. Moritz, and B. Mable (Sunderland, Mass.: Sinauer, 1996).

Protein electrophoresis. Every tissue (such as blood, muscle, or leaf) in each organism contains many different types of proteins specified by the thousands of structural genes within its cells. The laboratory challenge is to unveil protein variation in a population or species in such a way as then to attribute or "score" the results in terms of underlying genetic variation in specifiable genes that encoded those proteins.

A small tissue sample from each animal or plant is minced in liquid buffer. The resulting homogenate contains a vast ensemble of different types of protein molecules. A drop of this liquid then is placed into one of about twenty slots in a slablike gel. These starting slots are at the heads of parallel columns (like lanes in a bowling alley) down which the proteins soon will migrate. An electrical current is applied, causing the still-invisible proteins to move down the lanes at varying speeds, depending on each molecule's inherent net electrical charge. Each lane carries the proteins from a different individual animal or plant.

After several hours, the electrical current is turned off and the gel is soaked in a biochemical stain that reveals, as purple bands visible to the naked eye, the positions in the gel to which the relevant proteins migrated (Figure A1). Each organism then is scored for its genetic makeup ("genotype") at the corresponding gene, or locus. In a typical study, such results are accumulated for about thirty to fifty different genes in perhaps many hundreds of individuals.

Protein assays are not as sensitive as some of the DNA methods to be described next, in part because only a fraction of DNA-level variation results in protein-level variation. Nonetheless, protein-electrophoresis is well suited for many biological problems, such as assessing patterns of gene movement in a species, or detecting hybridization events (crosses) between species.

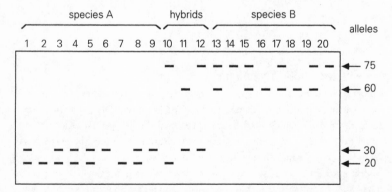

FIGURE A1. Schematic representation of a protein-electrophoretic gel with 20 lanes, each representing a different specimen. Shown are band profiles for a protein specified by a single gene, in this case with four different forms ("alleles," designated by numbers at the right). In species A, individuals 2 and 7 possess only (are "homozygous" for) the "20" allele, individuals 6 and 9 are homozygous for the "30" allele, and the other individuals are "heterozygous" (display two different alleles, in this case "30" and "20"). Likewise, the genotype of each specimen in species B can be scored as homozygous or heterozygous for the "60" and "75" alleles. Note that any first-generation hybrid (lanes 10–12) would carry one allele from each of the two species.

DNA fingerprinting. This term refers to a class of DNA-level assays that typically distinguish each individual plant or animal, much the way that conventional fingerprints uniquely mark each person. This power stems from the highly variable nature of "minisatellite" or "microsatellite" DNA regions.

A "minisatellite" is a stretch of DNA containing a series of short repetitive sequence units. Some of these units are tandemly arranged; others are scattered about the genome. The number and arrangement of units differ from one individual to the next. Thus, when a minisatellite complex is passed though a gel (by laboratory procedures resembling protein electrophoresis), the molecules separate in complicated patterns unique to each specimen, yielding ornate gel profiles (Figure A2). Like

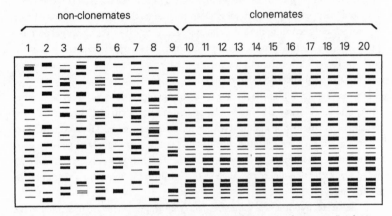

FIGURE A2. Schematic representation of a minisatellite "DNA fingerprint" gel. Note how these complex banding patterns resemble bar codes, and also how only clonemates share an identical DNA profile.

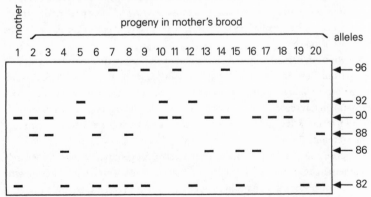

FIGURE A3. Schematic representation of a microsatellite gel, presented here in the context of paternity analysis. The individual in the leftmost lane is the known mother of the 19 progeny in lanes to the right. In each offspring, one allele (either "82" or "90") is of maternal origin, so the other allele must have come from the father. Collectively, there are four alleles ("86," "88," "92," and "96") of paternal origin in the progeny, so the brood was sired by at least two different males.

conventional human fingerprints, these individual-specific "DNA fingerprints" have wonderful forensic power, especially for distinguishing clonemates (genetically identical individuals) from non-clonemates, and for assessing genetic parentage.

A "microsatellite" region is similar except that each repeated unit of DNA is even shorter. The assay method again involves electrophoresis, but in this case it focuses on one microsatellite locus at a time. The band profiles on gels (Figure A3) are interpreted much like those in protein electrophoresis (Figure A1), and they offer excellent power for clone identification and parentage analysis.

Restriction analysis. This method involves chopping up particular genes with molecular "scissors" called restriction enzymes, each of which cuts DNA at specific recognition sites. Typically, a gene of interest (such as mtDNA) is isolated and then "digested" with the enzyme, yielding DNA fragments that are then separated electrophoretically according to molecular size. In many cases, the DNA gel patterns themselves are the final product of the assays, used to identify different forms of the gene. Or the DNA "restriction fragments" may be subject to more refined assays, such as DNA sequencing, described next.

DNA sequencing. With suitable effort, DNA sequences (the actual precise order of DNA's building blocks—the nucleotides adenine [A], guanine [G], thymine [T], and cytosine [C]) can now be obtained readily for nearly any gene. The laboratory procedures are rather complicated, but basically they again involve fine-scale electrophoretic separations of DNA molecules. The resolving power is so high that each and every nucleotide difference (such as the substitution of a T for a G) can be detected readily.

To illustrate, hypothetical sequence data from three species are presented in the upper part of Figure A4. (Most real data sets would include hundreds or thousands of nucleotides from each of dozens to hundreds of specimens.) By studying the arrows in Figure A4, note that the sequences from species A and B differ from one another at only two positions (5 percent) along the string of forty nucleotide sites assayed, whereas their sequences both differ from that of species C at six nucleotide positions (15 percent).

The bottom part of Figure A4 shows an evolutionary tree, or phylogeny, estimated from these data, assuming that the DNA sequences evolved in clocklike fashion (in linear relation to time, an assumption that often can be tested critically with real data). In this representation, species A and B shared a common ancestor (at node D) more recently than the lineage leading to these two species shared a common ancestor (at node E) with C.

Suppose now that there is independent fossil or biogeographic evidence securely documenting that node E occurred 7.5 million years ago. This would imply that the common ancestor of A and B lived about 2.5 million years before the present. It would also imply that the rate of sequence evolution in this gene was about 1 percent *per lineage* per million years. The broader point is that once an approximate "molecular clock" for a given gene is calibrated against external evidence (for example, from fossils or other biogeographic information), it can be employed in turn to estimate separation times for taxa in which such independent evidence is lacking.

Polymerase chain reaction (PCR). This technique, introduced in the late 1980s, has

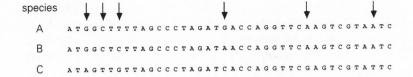

species

A   A T G G C T T T T A G C C C T A G A T G A C C A G G T T C A A G T C G T A A T C

B   A T G G C T C T T A G C C C T A G A T A A C C A G G T T C A A G T C G T A A T C

C   A T A G T T G T T A G C C C T A G A T C A C C A G G T T C G A G T C G T A T T C

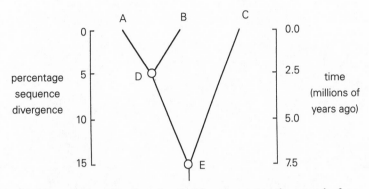

Figure A4. Upper section: Hypothetical nucleotide sequences in a short stretch of DNA from three extant species, with arrows indicating the positions displaying sequence variation. From such DNA sequence data, a phylogenetic tree (lower section) can be estimated (see text).

revolutionized the field of molecular genetics. Through a rapid series of biochemical steps in a test tube, the PCR method amplifies large numbers of perfect DNA copies from an original template of nucleotide sequence.

The starting biological material for PCR can be as little as a single drop of blood, a feather, a few grains of pollen, or a strand of hair. This means that geneticists can obtain suitable DNA samples without having to kill the organism, an important consideration especially when dealing with rare or endangered species. In some circumstances, PCR can amplify DNA even from well-preserved fossil material up to tens of thousands of years old. Once the DNA from a particular locus has been amplified by PCR, it then can be subjected to one or another of the DNA-level methods already mentioned (such as restriction analysis or sequencing).

DNA hybridization. Several additional molecular genetic assays are also available. One interesting method is atypical in that it does not involve an electrophoretic step. Instead, DNA strands are "hybridized."

Native DNA is a "double helix" consisting of two complementary strands held together by molecular bonds between properly paired nucleotides. DNA strands from different species normally differ from one another at varying fractions of nucleotide positions. Thus, when nucleotide strings from two species are forced to wed one another artificially in a test tube, the bindings won't be as firm as they are for double helical DNAs from the same species.

In the DNA hybridization approach, the difference in the tightness of binding between same-species ("homospecific") and different-species ("heterospecific") DNAs is quantified by comparing the molecules' "thermal stabilities." In a test tube, the temperature is increased gradually and the homo- and heterospecific double-helices are monitored for when they "melt." Results provide a useful measure of the magnitude of nucleotide divergence in these species' DNA sequences.

Glossary

ADAPTATION. Any feature (for example, morphological, physiological, behavioral) that helps an organism to survive and reproduce in a particular environment.

ADAPTIVE RADIATION. The rapid proliferation of related species as they evolve adaptations to different environmental conditions.

ALLELE. Any of the possible alternative forms of a given gene. A diploid individual carries two alleles at each autosomal gene, and these can be either identical in state (in which case the gene is homozygous) or different in state (heterozygous).

ALLOPARENTAL CARE (OF YOUNG). The rearing of foster children.

ALLOPATRIC. Inhabiting different geographic areas.

ALTRUISM. Selfless behavior performed for the benefit of others.

AMINO ACID. One of the molecular subunits that when joined together form a polypeptide.

AMPHIDROMY. An aquatic life cycle in which an individual spends most of its life in freshwater, except for a brief marine excursion during the larval stage.

ANADROMY. An aquatic life cycle in which an individual migrates from saltwater to freshwater to reproduce.

ANDRODIOECY. A condition in which hermaphrodites and males both exist within a plant species.

APOSEMATIC COLORATION. *See* warning coloration.

ASEXUAL REPRODUCTION. Any form of reproduction that does not involve the fusion of sex cells (gametes).

AUTOSOME. A chromosome in the nucleus other than a sex chromosome; in diploid organisms, autosomes are present in homologous pairs.

BACKCROSS. *n:* the progeny from a cross between an offspring and one of its parents; *v:* to conduct such a mating.

BACTERIUM. A unicellular microorganism without a true cellular nucleus.

BROOD PARASITISM. The use of a foster parent to raise an individual's young. The foster parent may be of the same or a different species (intraspecific and interspecific brood parasitism, respectively).

CANNIBALISM. Eating the flesh of one's own species.

CARNIVOROUS. Flesh-eating.

CATADROMY. An aquatic life cycle in which an individual migrates from freshwater to saltwater to reproduce.

CELL. A small, membrane bound unit of life capable of self-reproduction.

CHIMERA. An individual composed of a mixture of genetically different cells.

CHLOROPLAST. An organelle in the cytoplasm of plant cells that contains its own DNA (cpDNA) and is the site of photosynthesis.

CHROMOSOME. A threadlike structure within a cell that carries genes.

CLADE. A group of species (or, sometimes, individuals) that share a closer common ancestry with one another than with any other such group.

CLONE. *n:* a group of genetically identical cells or organisms, all descended from a single ancestral cell or organism; *v:* to produce such genetically identical cells or organisms.

COALESCENT THEORY. The body of mathematical thought concerning how alleles in any population trace back through the pedigree to common ancestral states in the past.

CONVERGENT EVOLUTION. The independent evolution of structural, functional, or other similarities between distantly related or unrelated species.

CUCKOLD. An individual whose mate has been unfaithful.

CUCKOLDRY. The act of making a cuckold.

CYTONUCLEAR ANALYSIS. The joint population-level examination of genes housed in cell nuclei and cytoplasms.

CYTOPLASM. The portion of a cell outside of the nucleus.

DEOXYRIBONUCLEIC ACID (DNA). The genetic material of most life forms; a double-stranded molecule composed of strings of nucleotides.

DIOECY. Used of plant species in which males and females are separate individuals.

DIPLOID. A usual condition of a somatic cell in which two copies of each chromosome are present.

DISPERSAL. Any spatial movement of an individual, usually away from its birth site or prior breeding location.

DIZYGOTIC TWINS (OR FRATERNAL TWINS). Genetically nonidentical siblings that stem from two separate zygotes during a pregnancy.

DNA/DNA HYBRIDIZATION. A class of laboratory procedures in which single-strand stretches of polynucleotides attract and bind to homologous, complementary single strands.

DNA FINGERPRINTING. A class of laboratory procedures in which individual-specific patterns of genetic variation are revealed via electrophoretic separation of multiple DNA fragments.

ECOLOGY. The study of the interrelationships among living organisms and their environments.

ECOSYSTEM. A community of ecologically interacting organisms and their environment.

EGG. A female gamete.

EGG MIMIC. A physical structure or pigment area on a male that closely resembles an egg in that species.

ELECTROPHORESIS. The movement of charged proteins or nucleic acids through a supporting gel under the influence of an electric current.

ENDANGERED SPECIES. A species at immediate risk of extinction.

ENDEMIC. Native to, and restricted to, a particular geographic area.

ENZYME. A protein that catalyzes a specific chemical reaction.

EUSOCIALITY. A social system characterized by cooperative care of young and reproductive division of labor with sterile individuals working on behalf of reproducers within a colony.

EVOLUTION. In simplest terms, any change across time in the genetic composition of populations or species.

EVOLUTIONARY TREE. *See* phylogeny.

EXOTIC. Not native to the geographic area in question.

EXTINCTION. The permanent disappearance of a population or species.

FERTILIZATION. The union of two gametes to produce a zygote.

FILIAL CANNIBALISM. Eating the flesh of one's own children.

FITNESS (GENETIC). The contribution of an individual (or of a particular genotype) to the next generation, relative to the contributions of other individuals (or genotypes) in the population.

FORENSICS (GENETIC). Of or pertaining to the diagnosis of otherwise unknown biological material based on analysis of proteins or DNA.

FOSSIL. Any remain or trace of life no longer alive.

FOUNDER EFFECT. Genetic consequences that follow the establishment of a new population from a small number of colonizing individuals.

GAMETE. A mature reproductive sex cell (egg or sperm).

GENE. The basic unit of heredity; usually taken to imply a sequence of nucleotides specifying production of a polypeptide or other functional product, but also can be applied to stretches of DNA with unknown or unspecified function.

GENEALOGY. A record of descent from ancestors through a pedigree.

GENE FLOW. The spatial movement of genes, normally within a species.

GENE POOL. The sum total of all hereditary material in a population or species.

GENET. All cells or individuals, however physically arranged, that stemmed from a single zygote.

GENETIC. Of or pertaining to the study of genes.

GLOSSARY

GENETIC DRIFT. Change in allele frequency in a finite population by chance sampling of gametes from generation to generation.

GENETIC ENGINEERING. The purposeful alteration of genetic material by humans.

GENETIC MARKERS. Natural gene-based tags that exist in all forms of life.

GENOME. The complete genetic constitution of an organism; also can refer to a particular composite piece of DNA, such as the mitochondrial genome.

GENOTYPE. The genetic constitution of an individual with reference to a single gene or set of genes.

GERM CELL. A sex cell or gamete.

GYNODIOECY. A condition in which hermaphrodites and females both exist within a plant species.

GYNOGENESIS. A form of reproduction related to parthenogenesis but in which an egg is stimulated to begin dividing by contact with sperm.

HAPLOID. A usual condition of a gametic cell in which one copy of each chromosome is present.

HERBIVOROUS. Plant-eating.

HEREDITY. Inheritance of genes; the phenomenon of familial transmission of genetic material from one generation to the next.

HERMAPHRODITE. A condition in which an individual produces both male and female gametes. If both types of gametes are generated at the same life stage, the individual is a synchronous hermaphrodite. If they are generated sequentially during an organism's lifetime, the individual is either a protandrous hermaphrodite (male gametes first) or a protogynous hermaphrodite (female gametes first).

HETEROGAMETIC SEX. The gender that produces gametes containing unlike sex chromosomes (in humans, the male).

HETEROZYGOTE. A diploid organism possessing two different alleles at a specified gene.

HOMOGAMETIC SEX. The gender that produces gametes containing sex chromosomes that are alike (in humans, the female).

HOMOLOGY. Similarity of features (morphological, molecular, and so on) due to inheritance from a shared ancestor.

HOMOZYGOTE. A diploid organism possessing two identical alleles at a specified gene.

HYBRIDIZATION. The successful mating of individuals belonging to genetically different populations or species.

HYBRIDOGENESIS. A quasi-sexual form of reproduction in which egg and sperm fuse to initiate embryonic development, but germ cells in the offspring later undergo an abnormal meiosis in which the resulting gametes carry no paternally derived genes.

INBREEDING. The mating of kin.

INBREEDING DEPRESSION. A loss in genetic fitness due to inbreeding.

INTROGRESSION. The movement of genes between species via hybridization and backcrossing.

INVERTEBRATE. An animal that does not possess a backbone.

JUMPING GENE. *See* transposable element.

JUNK DNA. A term formerly used to describe nucleic acid sequences that do not specify a functional protein or RNA product. *See also* selfish DNA.

KIN SELECTION. A form of natural selection due to individuals favoring the survival and reproduction of genetic relatives other than offspring.

LARVA (PL. LARVAE). The distinctive, preadult form in which some animals hatch from the egg.

LEK. A specific location where males of a species aggregate for courtship display.

LIFE CYCLE. The sequence of events for an individual, from its origin as a zygote to its death; one generation.

LOCUS (PL. LOCI). A gene.

MATERNITY ANALYSIS. *See* parentage analysis.

MATING SYSTEM. The particular pattern by which males and females, or their gametes, come together during the reproductive process. *See* monogamy, polyandry, polygamy, polygynandry, polygyny, promiscuity.

MATRILINE. A genetic transmission pathway strictly through females (as traversed, for example by animal mtDNA).

MEIOSIS. The cellular process whereby a diploid cell divides to form haploid gametes.

METABOLISM. The sum of all physical and chemical processes by which living matter is produced and maintained, and by which cellular energy is made available.

MICROBE. A very small organism visible only under a microscope.

MICROSATELLITE LOCUS. A stretch of DNA containing short repeated sequences each typically two, three, or four base-pairs in length.

MIGRATION. A periodic, typically seasonal movement to and from a given geographical area, often along a consistent route.

MIMICRY. The evolution of a close resemblance between any two unrelated species to deceive a third.

MINISATELLITE LOCUS. A stretch of DNA containing repeated sequences each typically about twenty to one hundred base-pairs in length.

MITOCHONDRION. An organelle in the cytoplasm of animal and plant cells that contains its own DNA (mtDNA) and is the site of some of the metabolic pathways involved in cellular energy production.

MITOSIS. A process of cell division that produces daughter cells with the same chromosomal constitution as the parent cell.

MOBILE ELEMENT. *See* transposable element.

MOLECULAR CLOCK. An evolutionary timepiece based on the evidence that genes or proteins tend to accumulate mutational differences at roughly constant rates in particular sets of lineages.

MOLECULAR MARKER. *See* genetic marker.

MONOECIOUS. The production of both male and female gametes by an individual plant.

MONOGAMY. A mating system in which one male is paired with one female.

MONOPHYLETIC. In evolution, tracing to a common or shared ancestor.

MONOZYGOTIC TWINS. Genetically identical siblings (barring mutation) that stem from a single zygote during a pregnancy.

MORPHOLOGY. The visible structures of organisms.

MOSAIC EVOLUTION. Variation in rates or patterns of evolution in different kinds of traits.

MUTATION. A change in the genetic constitution of an organism.

NATAL HOMING. Returning to the site of one's birth.

NATURAL HISTORY. The study of nature and natural phenomena.

NATURAL SELECTION. The differential contribution by individuals of different genotypes to the population of offspring in the next generation.

NEPOTISM. Favoritism directed toward one's genetic kin.

NUCLEIC ACID. *See* deoxyribonucleic acid.

NUCLEOTIDE. A unit of DNA consisting of a nitrogenous base, a pentose sugar, and a phosphate group.

NUCLEUS (PL. NUCLEI). The portion of a cell bounded by a nuclear membrane and containing chromosomes.

ORGANELLE. A complex, recognizable structure in the cell cytoplasm (such as a mitochondrion or chloroplast).

OUTCROSSING. The mating of unrelated individuals.

PAEDOPHAGE. An animal that feeds on the embryos or young of other species.

PALEONTOLOGY. The study of extinct forms of life, normally through fossils.

PARASITE. An organism that at some time in its life cycle is intimately associated with and harmful to a host.

PARENTAGE ANALYSIS (GENETIC). The assignment or estimation of maternity or paternity based on molecular markers.

PARTHENOGENESIS. The development of an individual from an unfertilized egg.

PATERNITY ANALYSIS. *See* parentage analysis.

PATHOGEN. An organism that produces a disease.

PEDIGREE. A diagram displaying population ancestry (mating partners and their offspring across generations).

PHOTOSYNTHESIS. The biochemical process by which a plant uses light to produce carbohydrates from carbon dioxide and water.

PHYLOGENETIC. Of or pertaining to phylogeny.

PHYLOGENY. Evolutionary relationships (historical descent) of a group of organisms or species.

PHYLOGEOGRAPHY. A field of scientific study concerned with the spatial distributions of genealogical lineages, including those within species.

PHYSIOLOGY. Study of the normal tissue operations and metabolic functions of living organisms.

PLANKTON. Small organisms that float freely in the ocean or other bodies of water.

PLEISTOCENE EPOCH. The geological time frame beginning about 2,000,000 years ago and ending roughly 10,000 years before the present.

POLLEN. A male gamete in plants.

POLLEN COMPETITION. In plants, rivalry among male gametes for fertilization success.

POLLINATION. The transfer of pollen to a female flower or flower part.

POLYANDRY. A mating system in which a female acquires and mates with multiple males, but a male typically has only one mate at most. *See also* polygamy and polygyny.

POLYEMBRYONY. The production of genetically identical offspring within a clutch or litter.

POLYGAMY. A mating system in which an individual has more than one mate. *See also* polygyny and polyandry.

POLYGYNANDRY. A mating system in which both males and females normally have several mates.

POLYGYNY. A mating system in which a male acquires and mates with multiple females, but a female typically has only one mate at most. *See also* polyandry and polygamy.

POLYMERASE CHAIN REACTION (PCR). A laboratory procedure for the in vitro replication of DNA from even small starting quantities.

POLYMORPHISM. With respect to particular organismal features or genotypes, the presence of two or more distinct forms in a population.

POLYPEPTIDE. A string of amino acids.

POLYPHYLETIC. In evolution, a group of organisms classified together, but tracing to different ancestral groups.

POPULATION. All individuals of a species normally inhabiting a defined area.

POPULATION BOTTLENECK. A severe but temporary reduction in the size of a population.

POPULATION STRUCTURE (GENETIC). Differences among geographic populations in genetic makeup.

PREDATOR. An organism that feeds by preying on other organisms.

PRIMER (FOR PCR). A short string of nucleotides used in conjunction with an appropriate enzyme to initiate synthesis of a nucleic acid.

PROMISCUITY. A mating system in which both males and females have many mates, or their gametes mix freely.

PROTEIN. A macromolecule composed of one or more polypeptide chains.

RAMET. A recognizable module or individual that may be part of a larger genet.

RECOMBINATION (GENETIC). The formation of new combinations of genes, as, for example, occurs naturally via meiosis and fertilization.

REGULATORY GENE. A gene that exerts operational control over the expression of other genes.

REPRODUCTIVE ISOLATION. Barriers to successful hybridization or introgression between biological species.

RESTRICTION ENZYME. An enzyme produced by a bacterium that cleaves foreign DNA molecules at specific recognition sites.

RESTRICTION FRAGMENT. A linear segment of DNA resulting from cleavage of a longer segment by a restriction enzyme.

RETROVIRUS. RNA viruses that utilize reverse transcription during their life cycle to integrate into the DNA of host cells.

RIBONUCLEIC ACID (RNA). The genetic material of many viruses, similar in structure to DNA. Also, any of a class of molecules that normally arise in cells from the transcription of DNA.

SELF-FERTILIZATION ("SELFING"). The union of male and female gametes from the same individual.

SELFISH DNA. DNA that displays self-perpetuating modes of behavior without apparent benefit to the organism. *See also* junk DNA.

SEX CHROMOSOME. A chromosome in the cell nucleus involved in distinguishing the two genders.

SEX RATIO. The relative numbers of males and females in a population.

SEXUAL DIMORPHISM. Marked differences in appearance between the males and females of a given species.

SEXUAL REPRODUCTION. Reproduction involving the production and subsequent fusion of haploid gametes.

SEXUAL SELECTION. The differential ability of individuals of the two genders to acquire mates. Intrasexual selection refers to competition among members of the same gender over access to mates; intersexual selection refers to patterns of mate choice by males and females.

SOMATIC CELL. Any cell in a multicellular organism other than those destined to become gametes.

SPECIATION (OR SPECIATION EVENT). The evolutionary formation of new species, often as a consequence of subdividing an ancestral species.

SPECIES (BIOLOGICAL). Groups of actually or potentially interbreeding individuals that are reproductively isolated from other such groups.

SPERM. A male gamete in animals.

SPERM COMPETITION. In animals, rivalry among male gametes for fertilization success.

SPERM STORAGE. The sequestering by females of male-derived sperm from past mating events.

SYMBIOSIS. Close associations between individuals of two or more species, sometimes but not necessarily restricted to mutually beneficial collaborations.

SYMPATRIC. Inhabiting the same geographic area.

SYSTEMATICS. The comparative study and classification of organisms particularly with regard to their phylogenetic relationships.

TAXON (PL. TAXA). A biotic lineage or entity deemed sufficiently distinct from other such lineages as to be worthy of a formal taxonomic name.

TAXONOMY. The practice of naming and classifying organisms. Hierarchical categories (from most to least inclusive) in common use are as follows: kingdom, phylum, class, order, family, tribe, genus, and species.

TRANSCRIPTION. The cellular process by which an RNA molecule is formed from a DNA template.

TRANSPOSABLE ELEMENT. Any of a class of DNA sequences that can move from one chromosomal site to another, often replicatively.

TRANSPOSON. *See* transposable element.

UNISEXUAL. Consisting of one sex only.

UTERUS. A mammalian organ in which embryos develop after implantation.

VERTEBRATE. An animal that possesses a backbone.

VICARIANCE. The process by which a historical barrier to dispersal can lead to the evolutionary emergence of two or more closely related forms of animals or plants in different geographical areas.

VIRUS. A tiny, obligate intracellular parasite, incapable of autonomous replication, that utilizes the host cell's replicative machinery.

WARNING COLORATION. Conspicuous coloration that serves to advertise the noxious, unpalatable, or otherwise dangerous properties of an organism to a potential predator.

X-CHROMOSOME. The sex chromosome normally present as two copies in female mammals (the homogametic sex) but as only one copy in males (the heterogametic sex).

Y-CHROMOSOME. In mammals, the sex chromosome normally present in males only.

ZYGOTE. Fertilized egg; the diploid cell arising from the union of male and female haploid gametes.

References and Further Reading

1 Some Evolutionary Oddities

THE PANDA'S PEDIGREE

O'Brien, S. J. 1987. The ancestry of the giant panda. *Scientific American* 257 (5): 102–107.

O'Brien, S. J., W. G. Nash, D. E. Wildt, M. E. Bush, and R. E. Benveniste. 1985. A molecular solution to the riddle of the giant panda's phylogeny. *Nature* 317:140–144.

Sarich, V. M. 1973. The giant panda is a bear. *Nature* 245:218–220.

MEAT-EATING PLANTS

Albert, V. A., S. E. Williams, and M. W. Chase. 1992. Carnivorous plants: phylogeny and structural evolution. *Science* 257:1491–1495.

Darwin, C. 1875. *Insectivorous Plants*. London: John Murray.

VENOMOUS VIPERS AND THEIR TOXINS

Daltry, J. C., W. Wüster, and R. S. Thorpe. 1996. Diet and snake venom evolution. *Nature* 379:537–540.

Minton, S. A., and S. R. Minton. 1969. *Venomous Reptiles*. New York: Scribner's.

GIANT TUBEWORMS

Black, M. B., K. M. Halanych, P. A. Y. Maas, W. R. Hoeh, J. Hashimoto, D. Desbruyeres, R. A. Lutz, and R. C. Vrijenhoek. 1997. Molecular systematics of vestimentiferan tubeworms from hydrothermal vents and cold-water seeps. *Marine Biology* 130:141–149.

Black, M. B., R. A. Lutz, and R. C. Vrijenhoek. 1994. Gene flow among vestimentiferan tube worm (*Riftia pachyptila*) populations from hydrothermal vents of the eastern Pacific. *Marine Biology* 120:33–39.

Grassle, J. F. 1985. Hydrothermal vent animals: distribution and biology. *Science* 229:713–717.

McHugh, D. 1997. Molecular evidence that echiurans and pogonophorans are derived annelids. *Proceedings of the National Academy of Sciences USA* 94:8006–8009.

HORSESHOE CRABS

Avise, J. C., W. S. Nelson, and H. Sugita. 1994. A speciational history of "living fossils": molecular evolutionary patterns in horseshoe crabs. *Evolution* 48:1986–2001.

Rudloe, A., and J. Rudloe. 1981. The changeless horseshoe crab. *National Geographic* 159:562–572.

Saunders, N. C., L. G. Kessler, and J. C. Avise. 1986. Genetic variation and geographic differentiation in mitochondrial DNA of the horseshoe crab, *Limulus polyphemus*. *Genetics* 112:613–627.

Selander, R. K., S. Y. Yang, R. C. Lewontin, and W. E. Johnson. 1970. Genetic variation in the horseshoe crab (*Limulus polyphemus*), a phylogenetic "relic." *Evolution* 24:402–414.

A TALE OF THE KING AND THE HERMIT

Cunningham, C. W., N. W. Blackstone, and L. W. Buss. 1992. Evolution of king crabs from hermit crab ancestors. *Nature* 355:539–542.

Gould, S. J. 1992. We are all monkey's uncles. *Natural History* 101 (6): 14–21.

PLANTLIKE ANIMALS FULL OF ALGAE

Glynn, P. W. 1993. Coral reef bleaching: ecological perspectives. *Coral Reefs* 12:1–17.

Rowan, R., N. Knowlton, A. Baker, and J. Jara. 1997. Landscape ecology of algal symbionts creates variation in episodes of coral bleaching. *Nature* 388:265–269.

Rowan, R., and D. A. Powers. 1992. Ribosomal RNA sequences and the diversity of symbiotic dinoflagellates (zooxanthellae). *Proceedings of the National Academy of Sciences USA* 89:3639–3643.

Taylor, F. J. R., ed. 1987. *The Biology of Dinoflagellates*. Oxford: Blackwell.

THE BACTERIAL BOUNTY WITHIN

Margulis, L. 1970. *Origin of Eukaryotic Cells*. New Haven, Conn.: Yale University Press.

Woese, C. R., A. Kandler, and M. L. Wheelis. 1990. Towards a natural system of organisms: proposal for the domains Archaea, Bacteria, and Eucarya. *Proceedings of the National Academy of Sciences USA* 87:4576–4579.

JUMPING GENES: NATURE'S REAL MOVERS AND SHAKERS

Avise, J. C. 1998. *The Genetic Gods: Evolution and Belief in Human Affairs*. Cambridge: Harvard University Press.

McDonald, J. F., ed. 1993. *Transposable Elements and Evolution*. Dordrecht: Kluwer Academic.

Ochert, A. 1999. Transposons. *Discover* 20 (12): 59–66.

Xiong, Y., and T. H. Eickbush. 1990. Origin and evolution of retroelements based upon their reverse transcriptase sequences. *EMBO Journal* 9:3353–3362.

2 Clones and Chimeras

NATURE'S CLONE-MAKING MAMMAL

Loughry, W. J., P. A. Prodöhl, C. M. McDonough, and J. C. Avise. 1998. Polyembryony in armadillos. *American Scientist* 86:274–279.

Prodöhl, P. A., W. J. Loughry, C. M. McDonough, W. S. Nelson, and J. C. Avise. 1996. Molecular documentation of polyembryony and the micro-spatial dispersion of clonal sibships in the nine-banded armadillo, *Dasypus novemcinctus. Proceedings of the Royal Society of London,* Series B, 263:1643–1649.

Wilmut, I., A. E. Schnieke, J. McWhir, A. J. Kind, and K. H. S. Campbell. 1997. Viable offspring derived from fetal and adult mammalian cells. *Nature* 385:810–813.

THE LIZARD THAT DISPENSED WITH SEX

Densmore, L. D., C. C. Moritz, J. W. Wright, and W. M. Brown. 1989. Mitochondrial-DNA analyses and the origin and relative age of parthenogenetic lizards (genus *Cnemidophorus*). IV. Nine *sexlineatus*-group unisexuals. *Evolution* 43:969–983.

Wright, J. W., C. Spolsky, and W. M. Brown. 1983. The origin of the parthenogenetic lizard *Cnemidophorus laredoensis* inferred from mitochondrial DNA analysis. *Herpetologica* 39:410–416.

AMAZON SEXUAL PARASITES

Avise, J. C., J. C. Trexler, J. Travis, and W. S. Nelson. 1991. *Poecilia mexicana* is the recent female parent of the unisexual fish *P. formosa. Evolution* 45:1530–1533.

Dawley, R. M., and J. P. Bogart, eds. 1989. *Ecology and Evolution of Unisexual Vertebrates.* Albany: New York State Museum.

White, M. J. D. 1978. Cytogenetics of the parthenogenetic grasshopper *Warramaba* (formerly *Moraba*) *virgo* and its bisexual relatives. III. Meiosis of male "synthetic *virgo*" individuals. *Chromosoma* 67:55–61.

THE 100,000-YEAR-OLD CLONE

Avise, J. C., J. M. Quattro, and R. C. Vrijenhoek. 1992. Molecular clones within organismal clones: mitochondrial DNA phylogenies and the evolutionary histories of unisexual vertebrates. *Evolutionary Biology* 26:225–246.

Quattro, J. M., J. C. Avise, and R. C. Vrijenhoek. 1992. An ancient clonal lineage in the fish genus *Poeciliopsis* (Atheriniformes: Poeciliidae). *Proceedings of the National Academy of Sciences USA* 89:348–352.

WHEN IS AN INDIVIDUAL NOT AN INDIVIDUAL?

Anzenberger, G. 1992. Monogamous mating systems and paternity in primates. Pp. 203–224 in *Paternity in Primates: Genetic Tests and Theories,* edited by R. D. Martin, A. F. Dixson, and E. J. Wickings. Basel: Karger.

Benirschke, K., J. M. Anderson, and L. E. Brownhill. 1962. Marrow chimerism in marmosets. *Science* 138:513–515.

Haig, D. 1999. What is a marmoset? *American Journal of Primatology* 49:285–296.

Signer, E. N., G. Anzenberger, and A. J. Jeffreys. 2000. Chimaeric and constitutive DNA fingerprints in the common marmoset (*Callithrix jacchus*). *Primates* 41:49–61.

CHIMERIC SEA SQUIRTS

Buss, L. W. 1982. Somatic cell parasitism and the evolution of somatic tissue compatibility. *Proceedings of the National Academy of Sciences USA* 79:5337–5341.

Grosberg, R. K., and J. F. Quinn. 1986. The genetic control and consequences of kin recognition by the larvae of a colonial marine invertebrate. *Nature* 322:456–459.

Pancer, Z., H. Gershon, and B. Rinkevich. 1995. Coexistence and possible parasitism of somatic and germ line cells in chimeras of the colonial urochodate *Botryllus schlosseri*. *Biological Bulletin* 189:106–112.

Sabbadin, A., and G. Zaniolo. 1979. Sexual differentiation and germ cell transfer in the colonial ascidian *Botryllus schlosseri*. *Journal of Experimental Zoology* 207:289–304.

THE STRANGLER FIG GANG

Thomson, J. D., E. A. Herre, J. L. Hamrick, and J. L. Stone. 1991. Genetic mosaics in strangler fig trees: implications for tropical conservation. *Science* 254:1214–1216.

3 Hermaphroditism

THE FISH THAT MATES WITH ITSELF

Laughlin, T. L., B. A. Lubinski, E.-H. Park, D. S. Taylor, and B. J. Turner. 1995. Clonal stability and mutation in the self-fertilizing hermaphroditic fish, *Rivulus marmoratus*. *Journal of Heredity* 86:399–402.

Turner, B. J., J. F. Elder Jr., T. F. Laughlin, W. P. Davis, and D. S. Taylor. 1992. Extreme clonal diversity and divergence in populations of a selfing hermaphroditic fish. *Proceedings of the National Academy of Sciences USA* 89:10643–10647.

See also Web site: http://www.bsi.vt.edu/rivmar/review.htm.

HOW SNAILS SOW THEIR OATS

Allard, R. W. 1975. The mating system and microevolution. *Genetics* 79:115–126.

Clegg, M. T., and R. W. Allard. 1972. Patterns of genetic differentiation in the slender wild oat species *Avena barbata*. *Proceedings of the National Academy of Sciences USA* 69:1820–1824.

Hamrick, J. L., and R. W. Allard. 1972. Microgeographical variation in allozyme frequencies in *Avena barbata*. *Proceedings of the National Academy of Sciences USA* 69:2100–2104.

Selander, R. K., and D. W. Kaufman. 1973. Self-fertilization and genetic population structure in a colonizing land snail. *Proceedings of the National Academy of Sciences USA* 70:1186–1190.

Selander, R. K., and R. O. Hudson. 1976. Animal population structure under close inbreeding: the land snail *Rumina* in southern France. *American Naturalist* 110:685–718.

BARRIERS TO SELF-POLLINATION

De Nettancourt, D. 1977. *Incompatibility in Angiosperms*. New York: Springer-Verlag.

Ioerger, T. R., A. G. Clark, and T.-H. Kao. 1990. Polymorphism at the self-incompatibility locus in Solanaceae predates speciation. *Proceedings of the National Academy of Sciences USA* 87:9732–9735.

Schemske, D. W., and R. Lande. 1985. The evolution of self-fertilization and inbreeding depression in plants. II. Empirical observations. *Evolution* 39:41–52.

MORE SEXUAL CONFUSION

Darwin, C. 1877. *The Different Forms of Flowers on Plants of the Same Species*. London: Murray.

Liston, A., L. H. Rieseberg, and T. S. Elias. 1990. Functional androdioecy in the flowering plant *Datisca glomerata*. *Nature* 343:641–642.

Swensen, S. W., J. N. Luthi, and L. H. Rieseberg. 1998. Datiscaceae revisited: monophyly and the sequence of breeding system evolution. *Systematic Botany* 23:157–169.

THE FISH THAT CHANGES ITS SEX

Chapman, R. W., G. R. Sedberry, C. C. Koenig, and B. M. Eleby. 1999. Stock identification of gag, *Mycteroperca microlepis*, along the southeast coast of the United States. *Marine Biotechnology* 1:137–146.

Polovina, J. J., and S. Ralston, eds. 1987. *Tropical Snappers and Groupers: Biology and Fisheries Management*. Boulder, Colo.: Westview Press.

4 Sex, Pregnancy, and Making Babies

THE ONUS OF PREGNANCY

Birkhead, T. R. 2000. *Promiscuity*. Cambridge: Harvard University Press.

Birkhead, T. R., and A. P. Møller, eds. 1998. *Sperm Competition and Sexual Selection*. London: Academic Press.

Chesser, R. K., M. W. Smith, and M. H. Smith. 1984. Biochemical genetics of mosquitofish. III. Incidence and significance of multiple insemination. *Genetica* 64:77–81.

Zane, L., W. S. Nelson, A. G. Jones, and J. C. Avise. 1999. Microsatellite assessment of multiple paternity in natural populations of a live-bearing fish, *Gambusia holbrooki*. *Journal of Evolutionary Biology* 12:61–69.

PSEUDO-NUPTIAL FLIGHTS IN PSEUDOSCORPIONS

Newcomer, S. D., J. A. Zeh, and D. W. Zeh. 1999. Genetic benefits enhance the reproductive success of polyandrous females. *Proceedings of the National Academy of Sciences USA* 96:10236–10241.

Zeh, D. W., J. A. Zeh, and E. Bermingham. 1997. Polyandrous, sperm-storing females: carriers of male genotypes through episodes of adverse selection. *Proceedings of the Royal Society of London*, Series B, 264:119–125.

Zeh, J. A., S. D. Newcomer, and D. W. Zeh. 1998. Polyandrous females discriminate against previous mates. *Proceedings of the National Academy of Sciences USA* 95:13732–13736.

Zeh, J. A., and D. W. Zeh. 1994. Last-male sperm precedence breaks down when females mate with three males. *Proceedings of the Royal Society of London*, Series B, 257:287–292.

MALE PREGNANCY

Jones, A. G., and J. C. Avise. 1997. Microsatellite analysis of maternity and the mating system in the Gulf pipefish *Syngnathus scovelli*, a species with male pregnancy and sex-role reversal. *Molecular Ecology* 6:203–213.

———. 2001. Mating systems and sexual selection in male-pregnant pipefishes and seahorses: insights from microsatellite-based studies of maternity. *Journal of Heredity* 92:212–219.

Kvarnemo, C., G. I. Moore, A. G. Jones, W. S. Nelson, and J. C. Avise. 2000. Monogamous pair-bonds and mate-switching in the Western Australian seahorse *Hippocampus subelongatus*. *Journal of Evolutionary Biology* 13:882–888.

Lourie, S. A., A. C. J. Vincent, and H. J. Hall. 1999. *Seahorses: An Identification Guide to the World's Species and Their Conservation*. London: Project Seahorse.

Vincent, A., I. Ahnesjö, A. Berglund, and G. Rosenqvist. 1992. Pipefishes and seahorses: are they all sex role reversed? *Trends in Ecology and Evolution* 7:237–241.

A BIRD THAT CHOOSES THE SEX OF ITS CHILDREN

Griffiths, R., and B. Tiwari. 1993. The isolation of molecular genetic markers for the identification of sex. *Proceedings of the National Academy of Sciences USA* 90:8324–8326.

Komdeur, J., S. Daan, J. Tinbergen, and C. Mateman. 1997. Extreme adaptive modification in sex ratio of the Seychelles warbler's eggs. *Nature* 385:522–525.

ROLY-POLY SEX RATIOS

O'Neill, S. L., A. A. Hoffmann, and J. H. Werren, eds. 1997. *Influential Passengers: Inherited Microorganisms and Arthropod Reproduction*. Oxford: Oxford University Press.

Rigaud, T., D. Bouchon, C. Souty-Grosset, and R. Raimond. 1999. Mitochondrial DNA polymorphism, sex ratio distorters and population genetics in the isopod *Armadillidium vulgare*. *Genetics* 152:1669–1677.

THE SOCIAL EQUALITY OF GULLS

Griffiths, R., M. C. Double, K. Orr, and R. J. G. Dawson. 1998. A DNA test to sex most birds. *Molecular Ecology* 7:1071–1075.

Millar, C. D., D. M. Lambert, A. R. Bellamy, P. M. Stapleton, and E. C. Young. 1992. Sex-specific restriction fragments and sex ratios revealed by DNA fingerprinting in the brown skua. *Journal of Heredity* 83:350–355.

EXTREME SOCIAL BEHAVIOR AND GENDER CONTROL

Hölldobler, B., and E. O. Wilson. 1990. *The Ants*. Cambridge: Harvard University Press, Belknap Press.

Hamilton, W. D. 1964. The genetical theory of social behavior. *Journal of Theoretical Biology* 7:1–52.

Page, R. E., Jr. 1986. Sperm utilization in social insects. *Annual Review of Entomology* 31:297–320.

Ross, K. G., and R. W. Matthews, eds. 1991. *The Social Biology of Wasps*. Ithaca, N.Y.: Cornell University Press, Comstock.

REPTILES WHOSE SEX IS TEMPERATURE DEPENDENT

Bull, J. J. 1983. *Evolution of Sex-Determining Mechanisms*. Menlo Park, Calif.: Benjamin/Cummings.

Shine, R. 1999. Why is sex determined by nest temperature in many reptiles? *Trends in Ecology and Evolution* 14:186–189.

5 Unusual Mating Practices

FATHERLY DEVOTION AND FEMALE IMPERSONATORS

Blumer, L. S. 1979. Male parental care in the bony fishes. *Quarterly Review of Biology* 54:149–161.

Colbourne, J. K., B. D. Neff, J. M. Wright, and M. R. Gross. 1996. DNA fingerprinting of bluegill sunfish (*Lepomis macrochirus*) using $(GT)_n$ microsatellites and its potential for assessment of mating success. *Canadian Journal of Fisheries and Aquatic Science* 53:342–349.

DeWoody, J. A., D. E. Fletcher, S. D. Wilkins, W. S. Nelson, and J. C. Avise. 1998. Molecular genetic dissection of spawning, parentage, and reproductive tactics in a population of redbreast sunfish, *Lepomis auritus. Evolution* 52:1802–1810.

Neff, B. D. 2001. Genetic paternity analysis and breeding success in bluegill sunfish (*Lepomis macrochirus*). *Journal of Heredity* 92:111–119.

Philipp, D. P., and M. R. Gross. 1994. Genetic evidence for cuckoldry in bluegill *Lepomis macrochirus. Molecular Ecology* 3:563–569.

FEMALE ACCOMPLICES OF MALE CUCKOLDRY

Oring, L. W., R. C. Fleischer, J. M. Reed, and K. E. Marsden. 1992. Cuckoldry through stored sperm in the sequentially polyandrous spotted sandpiper. *Nature* 359:631–633.

LIZARDS THAT PLAY ROCK-PAPER-SCISSORS

Sinervo, B., and C. M. Lively. 1996. The rock-paper-scissors game and the evolution of alternative male strategies. *Nature* 380:240–243.

Zamudio, K. R., and B. Sinervo. 2000. Polygyny, mate-guarding, and posthumous fertilization as alternative male strategies. *Proceedings of the National Academy of Sciences USA* 97:14427–14432.

BIRDS WITH ROVING EYES AND CHEATING HEARTS

Avise, J. C. 1996. Three fundamental contributions of molecular genetics to avian ecology and evolution. *Ibis* 138:16–25.

Burke, T., N. B. Davies, M. W. Bruford, and B. J. Hatchwell. 1989. Parental care and mating behaviour of polyandrous dunnocks *Prunella modularis* related to paternity by DNA fingerprinting. *Nature* 338:249–251.

Davies, N. B. 1992. *Dunnock Behaviour and Social Evolution*. Oxford: Oxford University Press.

Gowaty, P. A., and W. C. Bridges. 1991. Nestbox availability affects extra-pair fertilizations and conspecific nest parasitism in eastern bluebirds, *Sialia sialis. Animal Behavior* 41:661–675.

Westneat, D. F., P. W. Sherman, and M. L. Morton. 1990. The ecology and evolution of extra-pair copulations in birds. Pp. 331–368 in *Current Ornithology*, vol. 7, edited by D. M. Power. New York: Plenum Press.

TREEFROG MATING CEREMONIES

Lamb, T., and J. C. Avise. 1986. Directional introgression of mitochondrial DNA in a hybrid population of tree frogs: the influence of mating behavior. *Proceedings of the National Academy of Sciences USA* 83:2526–2530.

Mattison, C. 1987. *Frogs and Toads of the World*. New York: Facts on File.

SWORDTAILS' TALES

Basolo, A. L. 1990. Female preference predates the evolution of the sword in sword-tail fish. *Science* 250:808–810.

———. 1995. Phylogenetic evidence for the role of a pre-existing bias in sexual selection. *Proceedings of the Royal Society of London,* Series B, 259:307–311.

Endler, J. A., and A. L. Basolo. 1998. Sensory ecology, receiver biases and sexual selection. *Trends in Ecology and Evolution* 13:415–420.

Meyer, A., J. M. Morrissey, and M. Schartl. 1994. Recurrent origin of a sexually selected trait in *Xiphophorus* fishes inferred from a molecular phylogeny. *Nature* 368:539–542.

WHY SOME SPECIES LIKE LEKS

Höglund, J., R. V. Alatalo, A. Lundberg, P. T. Rintamäki, and J. Lindell. 1999. Microsatellite markers reveal the potential for kin selection on black grouse leks. *Proceedings of the Royal Society of London,* Series B, 266:813–816.

Petrie, M., A. Krupa, and T. Burke. 1999. Peacocks lek with relatives even in the absence of social and environmental cues. *Nature* 401:155–157.

Sherman, P. W. 1999. Birds of a feather lek together. *Nature* 401:119–120.

THE NAKED MOLE RAT

Reeve, H. K., D. F. Westneat, W. A. Noon, P. W. Sherman, and C. F. Aquadro. 1990. DNA "fingerprinting" reveals high levels of inbreeding in colonies of the eusocial naked mole-rat. *Proceedings of the National Academy of Sciences USA* 87:2496–2500.

Sherman, P. W., J. U. M. Jarvis, and R. D. Alexander, eds. 1991. *The Biology of the Naked Mole-Rat.* Princeton, N.J.: Princeton University Press.

6 Novel Ways of Handling Eggs and Sperm

EGG-DUMPING AND WILY CUCKOOS

Gibbs, H. L., M. D. Sorenson, K. Marchetti, M. de L. Brooke, N. B. Davies, and H. Nakamura. 2000. Genetic evidence for female host-specific races of the common cuckoo. *Nature* 407:183–186.

Marchetti, K., H. Nakamura, and H. L. Gibbs. 1998. Host-race formation in the common cuckoo. *Science* 282:471–472.

Petrie, M., and A. P. Møller. 1991. Laying eggs in others' nests: intraspecific brood parasitism in birds. *Trends in Ecology and Evolution* 6:315–320.

Rothstein, S. I., and S. Robinson, eds. 1998. *Parasitic Birds and Their Hosts: Studies in Coevolution.* New York: Oxford University Press.

Westneat, D. F., and M. S. Webster. 1994. Molecular analyses of kinship in birds: interesting questions and useful techniques. Pp. 91–126 in *Molecular Ecology and Evolution: Approaches and Applications,* edited by B. Scheirwater, B. Streit, G. P. Wagner, and R. DeSalle. Basel: Birkhauser.

THE NEST ARCHITECTURE OF SWALLOWS

Collias, N. E., and E. C. Collias. 1984. *Nest Building and Bird Behavior.* Princeton, N.J.: Princeton University Press.

Turner, A., and C. A. Rose. 1989. *Handbook to the Swallows and Martins of the World.* London: Christopher Helm.

Winkler, D. W., and F. H. Sheldon. 1993. Evolution of nest construction in swallows (Hirundinidae): a molecular phylogenetic perspective. *Proceedings of the National Academy of Sciences USA* 90:5705–5707.

EGG THIEVERY AND NEST PIRACY

DeWoody, J. A., and J. C. Avise. 2001. Genetic perspectives on the natural history of fish mating systems. *Journal of Heredity* 92:167–172.

Jones, A. G., S. Ostlund-Nilsson, and J. C. Avise. 1998. A microsatellite assessment of sneaked fertilizations and egg thievery in the fifteenspine stickleback. *Evolution* 52:848–858.

Ridley, M., and C. Retchen. 1981. Female sticklebacks prefer to spawn with males whose nests contain eggs. *Behaviour* 76:152–161.

Taborsky, M. 1994. Sneakers, satellites, and helpers: parasitic and cooperative behavior in fish reproduction. *Advanced Studies in Behavior* 23:1–100.

MALES WHOSE BODY PARTS MIMIC EGGS

Knapp, R. A., and R. C. Sargent. 1989. Egg-mimicry as a mating strategy in the fantail darter, *Etheostoma flabellare*: females prefer males with eggs. *Behavioral Ecology and Sociobiology* 25:321–326.

Page, L. M., and H. L. Bart. 1989. Egg mimics in darters (Pisces: Percidae). *Copeia* 1989:514–518.

Porter, B. A., A. C. Fiumera, and J. C. Avise. 2002. Egg mimicry and allopaternal care: two mate attracting tactics by which nesting striped darter (*Etheostoma virgatum*) males enhance reproductive success. *Behavioral Ecology and Sociobiology* 51: 350–359.

THE STORAGE OF SPERM BY FEMALES

Birkhead, T. R., and A. P. Møller. 1993. Sexual selection and the temporal separation of reproductive events: sperm storage data from reptiles, birds, and mammals. *Biological Journal of the Linnean Society* 50:295–311.

Pearse, D. E., and J. C. Avise. 2001. Turtle mating systems: behavior, sperm storage, and genetic paternity. *Journal of Heredity* 92:206–211.

Pearse, D. E., F. J. Janzen, and J. C. Avise. 2001. Genetic markers substantiate long-term storage and utilization of sperm by female painted turtles. *Heredity* 86:378–384.

Smith, R. L., ed. 1984. *Sperm Competition and the Evolution of Animal Mating Systems.* New York: Academic Press.

DAMSELS AND DRAGONS

Cooper, C. G., P. L. Miller, and P. W. H. Holland. 1996. Molecular genetic analysis of sperm competition in the damselfly *Ischnura elegans* (Vander Linden). *Proceedings of the Royal Society of London,* Series B, 263:1343–1349.

Corbet, P. S. 1999. *Dragonflies: Behaviour and Ecology of Odonota.* Colchester, England: Harley Books.

Siva-Jothy, M. T., and R. E. Hooper. 1996. Differential use of stored sperm during oviposition in the damselfly *Calopteryx splendens xanthosoma* (Charpentier). *Behavioral Ecology and Sociobiology* 39:389–393.

Waage, J. K. 1979. Dual function of the damselfly penis: sperm removal and transfer. *Science* 203:916–918.

BEAUTIFUL IRIS

Carney, S. E., S. A. Hodges, and M. L. Arnold. 1996. Effects of differential pollen-tube growth on hybridization in the Louisiana irises. *Evolution* 50:1871–1878.

Emms, S. K., S. A. Hodges, and M. L. Arnold. 1996. Pollen-tube competition, siring success, and consistent asymmetric hybridization in Louisiana irises. *Evolution* 50:2201–2206.

EATING ONE'S OWN KIDS

DeWoody, J. A., D. E. Fletcher, S. D. Wilkins, and J. C. Avise. 2001. Genetic documentation of filial cannibalism in nature. *Proceedings of the National Academy of Sciences USA* 98:5090–5092.

Elgar, M. A., and B. J. Crespi, eds. 1992. *Cannibalism: Ecology and Evolution among Diverse Taxa.* Oxford: Oxford University Press.

Salfert, I. G., and E. E. Moodie. 1985. Filial cannibalism in the brook stickleback, *Culaea inconstans* (Kirtland). *Behaviour* 93:82–100.

7 Dispersal and Migration

THE TALE OF MOTHER GOOSE

Avise, J. C., R. T. Alisauskas, W. S. Nelson, and C. D. Ankney. 1992. Matriarchal population genetic structure in an avian species with female natal philopatry. *Evolution* 46:1084–1096.

Cooke, F. 1987. Lesser snow goose: A long-term population study. Pp. 407–432 in *Avian Genetics*, edited by F. Cooke and P. A. Buckley. New York: Academic Press.

Quinn, T. W. 1992. The genetic legacy of mother goose—phylogeographic patterns of lesser snow goose *Chen caerulescens* maternal lineages. *Molecular Ecology* 1:105–117.

THE FISH THAT BRAVES THE BERMUDA TRIANGLE

Avise, J. C., G. S. Helfman, N. C. Saunders, and L. S. Hales. 1986. Mitochondrial DNA differentiation in North Atlantic eels: population genetic consequences of an unusual life history pattern. *Proceedings of the National Academy of Sciences USA* 83:4350–4354.

Williams, G. C., and R. K. Koehn. 1984. Population genetics of North Atlantic catadromous eels (*Anguilla*). Pp. 529–560 in *Evolutionary Genetics of Fishes*, edited by B. J. Turner. New York: Plenum Press.

Wirth, T., and L. Bernatchez. 2001. Genetic evidence against panmixia in the European eel. *Nature* 409:1037–1040.

THE MIGRATORY CIRCUIT OF A WHALE

Baker, C. S., and S. R. Palumbi. 1996. Population structure, molecular systematics, and forensic identification of whales and dolphins. Pp. 10–49 in *Conservation Genetics: Case Histories from Nature*, edited by J. C. Avise and J. L. Hamrick. New York: Chapman and Hall.

Baker, C. S., S. R. Palumbi, R. H. Lambertsen, M. T. Weinrich, J. Calambokidis, and S. J. O'Brien. 1990. Influence of seasonal migration on geographic distribution of mitochondrial DNA haplotypes in humpback whales. *Nature* 344:238–240.

A SALUTE TO SALMON

Allendorf, F. W., and R. S. Waples. 1996. Conservation and genetics of salmonid fishes. Pp. 238–280 in *Conservation Genetics: Case Histories from Nature*, edited by J. C. Avise and J. L. Hamrick. New York: Chapman and Hall.

Levin, P. S., and M. H. Schiewe. 2001. Preserving salmon biodiversity. *American Scientist* 89:220–227.

Magnuson, J. J., and others. 1996. *Upstream: Salmon and Society in the Pacific Northwest.* Washington, D.C.: National Research Council.

Park, L. K., M. A. Brainard, D. A. Dightman, and G. A. Winans. 1993. Low levels of intraspecific variation in the mitochondrial DNA of chum salmon *(Oncorhynchus keta). Molecular Marine Biology and Biotechnology* 2:362–370.

THE KING OF MIGRATION

Ackery, P. R., and R. I. Vane-Wright. 1984. *Milkweed Butterflies.* Ithaca, N.Y.: Cornell University Press.

Brower, A. V. Z., and T. M. Boyce. 1991. Mitochondrial DNA variation in monarch butterflies. *Evolution* 45:1281–1286.

Eanes, W. F., and R. K. Koehn. 1978. An analysis of genetic structure in the monarch butterfly, *Danaus plexippus* L. *Evolution* 32:784–797.

Gibbs, G. 1994. *The Monarch Butterfly.* Auckland, New Zealand: Reed Publishing.

GREEN TURTLE ODYSSEYS

Bowen, B. W., and J. C. Avise. 1996. Conservation genetics of marine turtles. Pp. 190–237 in *Conservation Genetics: Case Histories from Nature*, edited by J. C. Avise and J. L. Hamrick. New York: Chapman and Hall.

Bowen, B. W., A. B. Meylan, and J. C. Avise. 1989. An odyssey of the green sea turtle: Ascension Island revisited. *Proceedings of the National Academy of Sciences USA* 86:573–576.

Carr, A., and P. J. Coleman 1974. Seafloor spreading theory and the odyssey of the green turtle from Brazil to Ascension Island, central Atlantic. *Nature* 249:128–130.

Meylan, A. B., B. W. Bowen, and J. C. Avise. 1990. A genetic test of the natal homing versus social facilitation models for green turtle migration. *Science* 248:724–727.

SWEET BEES WITH A NASTY DISPOSITION

Hall, H. G., and K. Muralidharan. 1989. Evidence from mitochondrial DNA that African honey bees spread as continuous maternal lineages. *Nature* 339:211–212.

Rinderer, T. E., J. A. Stelzer, B. P. Oldroyd, S. M. Buco, and W. L. Rubink. 1991. Hybridization between European and Africanized honey bees in the Neotropical Yucutan peninsula. *Science* 253:309–311.

Ruttner, F. 1988. *Biogeography and Taxonomy of Honey Bees.* Berlin: Springer-Verlag.

Sheppard, W. S., T. E. Rinderer, J. A. Mazzoli, J. A. Stelzer, and H. Shimanuki. 1991. Gene flow between African- and European-derived honey bee populations in Argentina. *Nature* 349:782–784.

Smith, D. R., O. R. Taylor, and W. M. Brown. 1989. Neotropical Africanized honey bees have African mitochondrial DNA. *Nature* 339:213–215.

THE BALLAST TRAVELERS

Carlton, J. T., and J. B. Geller. 1993. Ecological roulette: the global transport of non-indigenous marine organisms. *Science* 261:78–82.

Geller, J. B., E. D. Walton, E. D. Grosholz, and G. M. Ruiz. 1997. Cryptic invasions of the crab *Carcinus* detected by molecular phylogeography. *Molecular Ecology* 6:901–906.

National Research Council. 1996. *Stemming the Tide: Controlling Introductions of Nonindigenous Species by Ships' Ballast Water.* Washington, D.C.: National Academy Press.

8 Island Life

DARWIN'S GALÁPAGOS FINCHES

Christie, D. M., R. A. Duncan, A. R. McBirney, M. A. Richards, W. M. White, K. S. Harpp, and C. G. Fox. 1992. Drowned islands downstream from the Galápagos hotspot imply extended speciation times. *Nature* 355:246–248.

Freeland, J. R., and P. T. Boag. 1999. The mitochondrial and nuclear genetic homogeneity of the phenotypically diverse Darwin's ground finches. *Evolution* 53:1553–1563.

Grant, P. R. 1999. *Ecology and Evolution of Darwin's Finches.* Princeton, N.J.: Princeton University Press.

Larson, E. J. 2001. *Evolution's Workshop: God and Science on the Galápagos Islands.* New York: Basic Books.

Weiner, J. 1995. *The Beak of the Finch: A Story of Evolution in Our Time.* New York: Vintage.

BEAUTIFUL FLIES

Beverly, S. M., and A. C. Wilson. 1985. Ancient origin for Hawaiian Drosophilinae inferred from protein comparisons. *Proceedings of the National Academy of Sciences USA* 82:4753–4757.

Carson, H. L., and K. Y. Kaneshiro. 1976. *Drosophila* of Hawaii: systematics and ecological genetics. *Annual Review of Ecology and Systematics* 7:311–345.

DeSalle, R. 1992. The origin and possible time of divergence of the Hawaiian Drosophilidae: evidence from DNA sequences. *Molecular Biology and Evolution* 9:905–916.

MORE EXOTIC BEAUTIES OF TROPICAL ISLES

Fleischer, R. C., C. E. McIntosh, and C. L. Tarr. 1998. Evolution on a volcanic conveyor belt: using phylogeographic reconstructions and K-Ar-based ages of the Hawaiian Islands to estimate molecular evolutionary rates. *Molecular Ecology* 7:533–545.

Freed, L. A., S. Conant, and R. C. Fleischer. 1987. Evolutionary ecology and radiation of Hawaiian passerine birds. *Trends in Ecology and Evolution* 2:196–203.

Johnson, N. K., J. A. Marten, and C. J. Ralph. 1989. Genetic evidence for the origin and relationships of Hawaiian honeycreepers (Aves: Fringillidae). *Condor* 91:379–396.

Tarr, C. L., and R. C. Fleischer. 1995. Evolutionary relationships of the Hawaiian honeycreepers (Aves, Drepanidinae). Pp. 147–159 in *Hawaiian Biogeography*, edited by W. L. Wagner and V. A. Funk. Washington, D.C.: Smithsonian Institution Press.

Baldwin, B. G., and R. H. Robichaux. 1995. Historical biogeography and ecology of the Hawaiian silversword alliance (Asteraceae). Pp. 259–287 in *Hawaiian Biogeography*, edited by W. L. Wagner and V. A. Funk. Washington, D.C.: Smithsonian Institution Press.

Carr, G. D. 1987. Beggar's ticks and tarweeds: masters of adaptive radiation. *Trends in Ecology and Evolution* 2:192–195.

Robichaux, R. H., G. D. Carr, M. Liebman, and R. W. Pearcy. 1990. Adaptive radiation of the Hawaiian silversword alliance (Compositae-Madiinae): ecological, morphological, and physiological diversity. *Annals of the Missouri Botanical Garden* 77:64–72.

THE CLINGING GOBY FISH

Chubb, A. L., R. M. Zink, and J. M. Fitzsimons. 1998. Patterns of mtDNA variation in Hawaiian freshwater fishes: the phylogeographic consequences of amphidromy. *Journal of Heredity* 89:8–16.

Fitzsimons, J. M., R. M. Zink, and R. T. Nishimoto. 1990. Genetic variation in the Hawaiian stream goby, *Lentipes concolor*. *Biochemical Systematics and Ecology* 18:81–83.

Zink, R. M., J. M. Fitzsimons, D. L. Dittmann, D. R. Reynolds, and R. T. Nishimoto. 1996. Evolutionary genetics of Hawaiian freshwater fish. *Copeia* 1996:330–335.

FABULOUS, FABLED FROGS

Myers, C. W., and J. W. Daly. 1983. Dart-poison frogs. *Scientific American* 248 (2): 120–133.

Summers, K., E. Bermingham, L. Weigt, S. McCafferty, and L. Dahlstrom. 1997. Phenotypic and genetic divergence in three species of dart-poison frogs with contrasting parental behavior. *Journal of Heredity* 88:8–13.

CARIBBEAN CRUISES

Avise, J. C. 2000. *Phylogeography: The History and Formation of Species*. Cambridge: Harvard University Press.

Hedges, S. B. 1996. Historical biogeography of West Indian vertebrates. *Annual Review of Ecology and Systematics* 27:163–196.

9 Species Proliferations

WARBLER WARDROBES

Avise, J. C., J. C. Patton, and C. F. Aquadro. 1980. Evolutionary genetics of birds: comparative molecular evolution in New World warblers and rodents. *Journal of Heredity* 71:303–310.

Baker, K. 1997. *Warblers of Europe, Asia, and North Africa*. Princeton, N.J.: Princeton University Press.

Curson, J., D. Quinn, and D. Beadle. 1994. *New World Warblers*. London: Croom Helm.

Lovette, I. J., and E. Bermingham. 1999. Explosive speciation in the New World *Dendroica* warblers. *Proceedings of the Royal Society of London*, Series B, 266:1629–1636.

Price, T., I. J. Lovette, E. Bermingham, H. L. Gibbs, and A. D. Richman. 2000. The imprint of history on communities of North American and Asian warblers. *American Naturalist* 156:354–367.

UNITY AND DIVERSITY IN THE WINGED AUSSIES

Sibley, C. G., and J. E. Ahlquist. 1986. Reconstructing bird phylogeny by comparing DNA's. *Scientific American* 254 (2): 82–93.

———. 1990. *Phylogeny and Classification of Birds.* New Haven, Conn.: Yale University Press.

EVOLUTIONARY TREES AND ELEPHANTS' TRUNKS

Cope, E. D. 1884. The Condylartha. *American Naturalist* 18:790–805.

De Jong, W. W., A. Zweers, and M. Goodman. 1981. Relationships of aardvark to elephants, hyraxes and sea cows from α-crystallin sequences. *Nature* 292:538–540.

Hedges, S. B. 2001. Afrotheria: plate tectonics meets genomics. *Proceedings of the National Academy of Sciences USA* 98:1–2.

Madsen, O., M. Scally, C. J. Douady, D. J. Kao, R. W. DeBry, R. Adkins, H. M. Amrine, M. J. Stanhope, W. W. de Jong, and M. S. Springer. 2001. Parallel adaptive radiations in two major clades of placental mammals. *Nature* 409:610–614.

Murphy, W. J., E. Eizirik, W. E. Johnson, Y. P. Zhang, O. A. Ryder, and S. J. O'Brien. 2001. Molecular phylogenetics and the origins of placental mammals. *Nature* 409:614–618.

Springer, M. S., G. C. Cleven, O. Madsen, W. W. de Jong, V.G. Waddell, H. M. Armine, and M. J. Stanhope. 1997. Endemic African mammals shake the evolutionary tree. *Nature* 388:61–64.

Van Dijk, M. A. M., O. Madsen, F. Catzeflis, M. J. Stanhope, W. W. de Jong, and M. Page. 2001. Protein sequence signatures support the African clade of mammals. *Proceedings of the National Academy of Sciences USA* 98:188–193.

BRILLIANT BUTTERFLIES

Brower, A. V. Z. 1994. Rapid morphological radiation and convergence among races of the butterfly *Heliconius erato* inferred from patterns of mitochondrial DNA evolution. *Proceedings of the National Academy of Sciences USA* 91:6491–6495.

Nijhout, H. F. 1991. *The Development and Evolution of Butterfly Wing Patterns.* Washington, D.C.: Smithsonian Institution Press.

FLOCKS OF AFRICAN FISHES

Echelle, A. A., and I. Kornfield, eds. 1984. *Evolution of Fish Species Flocks.* Orono: University of Maine Press.

Fryer, G., and T. D. Iles. 1972. *The Cichlid Fishes of the Great Lakes of Africa.* Neptune City, N.J.: TFH Publishers.

Greenwood, P. H. 1981. *The Haplochromine Fishes of the East African Lakes.* Ithaca, N.Y.: Cornell University Press.

Meyer, A., T. D. Kocher, P. Basasibwaki, and A. C. Wilson. 1990. Monophyletic origin of Lake Victoria cichlid fishes suggested by mitochondrial DNA sequences. *Nature* 347:550–553.

Schliewen, U. K., D. Tautz, and S. Pääbo. 1994. Sympatric speciation suggested by monophyly of crater lake cichlids. *Nature* 368:629–632.

Seehausen, O., J. J. M. van Alphen, and F. Witte. 1997. Cichlid fish diversity threatened by eutrophication that curbs sexual selection. *Science* 277:1808–1811.

Stiassny, M. L. J., and A. Meyer. 1999. Cichlids of the rift lakes. *Scientific American* 280 (2): 64–69.

MICROBATS AND MEGABATS

Adkins, R. M., and R. L. Honeycutt. 1991. Molecular phylogeny of the superorder Archonta. *Proceedings of the National Academy of Sciences USA* 88:10317–10321.

Bailey, W. J., J. L. Slightom, and M. Goodman. 1992. Rejection of the "flying primate" hypothesis by phylogenetic evidence from the ε-globin gene. *Science* 256:86–89.

Mindell, D. P., C. W. Dick, and R. J. Baker. 1991. Phylogenetic relationships among megabats, microbats, and primates. *Proceedings of the National Academy of Sciences USA* 88:10322–10326.

Pettigrew, J. D. 1986. Flying primates? Megabats have the advanced pathway from eye to midbrain. *Science* 231:1304–1306.

ANOMALIES AND PARADOXES IN SUNFLOWERS

Grant, V. 1981. *Plant Speciation*. New York: Columbia University Press.

Rieseberg, L. H. 1997. Hybrid origins of plant species. *Annual Review of Ecology and Systematics* 28:359–389.

Rieseberg, L. H., R. Carter, and S. Zona. 1990. Molecular tests of the hypothesized hybrid origin of two diploid *Helianthus* species (Asteraceae). *Evolution* 44:1498–1511.

Rieseberg, L. H., C. Van Fossen, and A. M. Desrochers. 1995. Hybrid speciation accompanied by genomic reorganization in wild sunflowers. *Nature* 375:313–316.

SNAPPING SHRIMPS

Duffy, J. E. 1996. Eusociality in a coral-reef shrimp. *Nature* 381:512–514.

———. 1996. Species boundaries, specialization, and the radiation of sponge-dwelling alpheid shrimp. *Biological Journal of the Linnean Society* 58:307–324.

Knowlton, N. 1993. Sibling species in the sea. *Annual Review of Ecology and Systematics* 24:189–216.

Knowlton, N., and L. A. Weigt. 1998. New dates and new rates for divergence across the Isthmus of Panama. *Proceedings of the Royal Society of London*, Series B, 265:2257–2263.

10 Wildlife Forensics and Conservation

THE PLIGHT OF THE WHALES

Baker, C. S., F. Cipriano, and S. R. Palumbi. 1996. Molecular genetic identification of whale and dolphin products from commercial markets in Korea and Japan. *Molecular Ecology* 5:671–685.

Baker, C. S., and S. R. Palumbi. 1994. Which whales are hunted? A molecular genetic approach to monitoring whaling. *Science* 265:1538–1539.

Bowen, B. W. 1999. A field born in conservation's cold war. *Trends in Ecology and Evolution* 15:1–3.

236

REFERENCES AND FURTHER READING

PINNIPED PENISES

Busch, B. C. 1985. *The War against Seals. A History of the North American Seal Fishery.* Montreal: McGill-Queens University Press.

Geist, V. 1988. How markets in wildlife meat and parts, and the sale of hunting privileges, jeopardize wildlife conservation. *Conservation Biology* 2:1–12.

Malik, S., P. J. Wilson, R. J. Smith, D. M. Lavigne, and B. N. White. 1997. Pinniped penises in trade: a molecular-genetic investigation. *Conservation Biology* 11:1365–1374.

Riedman, M. 1990. *The Pinnipeds: Seals, Sea Lions, and Walruses.* Berkeley: University of California Press.

A TASTY TURTLE

Roman, J., and B. W. Bowen. 2000. The mock turtle syndrome: genetic identification of turtle meat purchased in the south-eastern United States of America. *Animal Conservation* 3:61–65.

Roman, J., S. D. Santhuff, P. E. Moler, and B. W. Bowen. 1999. Population structure and cryptic evolutionary units in the alligator snapping turtle. *Conservation Biology* 13:135–142.

"CAVIAR EMPTOR"

Birstein, V. J., P. Doukakis, B. Sorkin, and R. DeSalle. 1998. Population aggregation analysis of three caviar-producing species of sturgeons and implications of the species identification of black caviar. *Conservation Biology* 12:766–775.

Birstein, V. J., J. R. Waldman, and W. E. Bemis, eds. 1997. *Sturgeon Biodiversity and Conservation.* Dordrecht: Kluwer Academic.

DeSalle, R., and V. J. Birstein. 1996. PCR identification of black caviar. *Nature* 381:197–198.

See also Web site: http://www.caviaremptor.org/.

AN ENDANGERED BIRD IN THE BELLY OF A SNAKE

Avise, J. C., and W. S. Nelson. 1989. Molecular genetic relationships of the extinct dusky seaside sparrow. *Science* 243:646–648.

National Research Council. 1995. *Science and the Endangered Species Act.* Washington, D.C.: National Academy Press.

Nelson, W. S., T. Dean, and J. C. Avise. 2000. Matrilineal history of the endangered Cape Sable seaside sparrow inferred from mitochondrial DNA polymorphism. *Molecular Ecology* 9:809–813.

Quay, T. L., J. B. Funderburg Jr., D. S. Lee, E. F. Potter, and C. S. Robbins, eds. 1983. *The Seaside Sparrow, Its Biology and Management.* Raleigh: North Carolina State Museum of Natural History.

THE RIDLEY RIDDLES

Bowen, B. W., A. M. Clark, F. A. Abreu-Grobois, A. Chaves, H. A. Reichart, and R. J. Ferl. 1998. Global phylogeography of the ridley sea turtles (*Lepidochelys* spp.) as inferred from mitochondrial DNA sequences. *Genetica* 101:179–189.

Bowen, B. W., A. B. Meylan, and J. C. Avise. 1991. Evolutionary distinctiveness of the endangered Kemp's ridley sea turtle. *Nature* 352:709–711.

Magnuson, J. J., and others. 1990. *Decline of the Sea Turtles.* Washington, D.C.: National Academy of Sciences.

LAKES, SWAMPS, AND GENE POOLS

Avise, J. C., P. C. Pierce, M. J. Van Den Avyle, M. H. Smith, W. S. Nelson, and M. A. Asmussen. 1997. Cytonuclear introgressive swamping and species turnover of bass after an introduction. *Journal of Heredity* 88:14–20.

Jacobs, J. 1998. *Bass Fishing in Georgia.* Atlanta: Peachtree.

THE FISH WHOSE BABIES GET ALL MIXED UP

Eschmeyer, W. N., and E. S. Herald. 1983. *A Field Guide to Pacific Coast Fishes of North America.* Boston: Houghton Mifflin.

Johns, G. C., and J. C. Avise. 1998. Tests for ancient species flocks based on molecular phylogenetic appraisals of *Sebastes* rockfishes and other marine fishes. *Evolution* 52:1135–1146.

Rocha-Olivares, A. 1998. Multiplex haplotype-specific PCR: a new approach for species identification of the early life stages of rockfishes of the species-rich genus *Sebastes* Cuvier. *Journal of Experimental Marine Biology and Ecology* 231:279–290.

11 Some Genetic World Records

THE 100-TON MUSHROOM

Smith, M. L., J. N. Bruhn, and J. B. Anderson. 1992. The fungus *Armillaria bulbosa* is among the largest and oldest living organisms. *Nature* 356:428–431.

Smith, M. L., L. C. Duchesne, J. N. Bruhn, and J. B. Anderson. 1990. Mitochondrial genetics in a natural population of the plant pathogen *Armillaria. Genetics* 126:575–582.

COPEPODS: NATURE'S MOST ABUNDANT ANIMAL?

Avise, J. C., R. M. Ball, and J. Arnold. 1988. Current versus historical population sizes in vertebrate species with high gene flow: a comparison based on mitochondrial DNA lineages and inbreeding theory for neutral mutations. *Molecular Biology and Evolution* 5:331–344.

Bucklin, A., and P. H. Wiebe. 1998. Low mitochondrial diversity and small effective population sizes of copepods *Calanus finmarchicus* and *Nannocalanus minor:* possible impact of climatic variation during recent glaciation. *Journal of Heredity* 89:383–392.

MATING CHAMPIONS OF THE INSECT WORLD

Cabrera-Mireles, C. 1999. *University of Florida Book of Insect Records*, Chap. 36. On line: http://gnv.ifas.ufl.edu/~tjw/chap36.htm.

Moritz, R. F. A., P. Kryger, G. Koeniger, N. Koeniger, A. Estoup, and S. Tingek. 1995. High degree of polyandry in *Apis dorsata* queens detected by DNA microsatellite variability. *Behavioral Ecology and Sociobiology.* 37:357–363.

Page, R. E., Jr. 1986. Sperm utilization in social insects. *Annual Review of Entomology* 31:297–320.

REFERENCES AND FURTHER READING

REFERENCES AND FURTHER READING

Tingek, S. 1995. High degree of polyandry in *Apis dorsata* queens detected by DNA microsatellite variability. *Behavioral Ecology and Sociobiology* 37:357–363.

LONESOME GEORGE: THE WORLD'S LONELIEST BEAST?

Caccone, A., J. P. Gibbs, V. Ketmaier, E. Suatoni, and J. R. Powell. 1999. Origin and evolutionary relationships of giant Galápagos tortoises. *Proceedings of the National Academy of Sciences USA* 96:13223–13228.

Darwin, C. 1839. *Journal of the Researches into the Geology and Natural History of Various Countries Visited by H.M.S. Beagle, under the Command of Captain Fitzroy, R.N., from 1832 to 1836*. London: Henry Colburn.

Tierney, J. 1995. Lonesome George of the Galápagos. *Science 85* 6 (5): 50–61.

LIFE'S EARLIEST FARMERS

Chapela, I. H., S. A. Rehner, T. R. Schultz, and U. G. Mueller. 1994. Evolutionary history of the symbiosis between fungus-growing ants and their fungi. *Science* 266:1691–1694.

Hinkle, G., J. K. Wetterer, T. R. Schultz, and M. L. Sogin. 1994. Phylogeny of the attine ant fungi based on analysis of small subunit ribosomal RNA gene sequences. *Science* 266:1695–1697.

Mueller, U. G., S. A. Rehner, and T. R. Schultz. 1998. The evolution of agriculture in ants. *Science* 281:2034–2038.

Wilson, E. O. 1971. *The Insect Societies*. Cambridge: Harvard University Press, Belknap Press.

HOW LOW DO ROOT TIPS GO?

Casper, B. B., and R. B. Jackson. 1997. Plant underground competition. *Annual Review of Ecology and Systematics* 28:545–570.

Jackson, R. B., L. A. Moore, W. A. Hoffman, W. T. Pockman, and C. R. Linder. 1999. Ecosystem rooting depth determined with caves and DNA. *Proceedings of the National Academy of Sciences USA* 96:11387–11392.

Linder, C. R., L. A. Moore, and R. B. Jackson. 2000. A universal molecular method for identifying underground plant parts to species. *Molecular Ecology* 9:1549–1559.

A FISH RETURNED FROM THE DEAD

Erdmann, M. V., R. L. Caldwell, and M. K. Moosa. 1998. Indonesian "king of the sea" discovered. *Nature* 395:335.

Holder, M. T., M. V. Erdmann, T. P. Wilcox, R. L. Caldwell, and D. M. Hillis. 1999. Two living species of coelacanths? *Proceedings of the National Academy of Sciences USA* 96:12616–12620.

Meyer, A. 1995. Molecular evidence on the origin of tetrapods and the relationships of the coelacanth. *Trends in Ecology and Evolution* 10:111–116.

Thomson, K. S. 1991. *Living Fossil: The Story of the Coelacanth*. New York: Norton.

Zardoya, R., and A. Meyer. 1997. The complete DNA sequence of the mitochondrial genome of a "living fossil," the coelacanth (*Latimeria chalumnae*). *Genetics* 146:995–1010.

THE WORLD'S MOST SHOCKING MARRIAGE?

Bowers, J., J.-M. Boursiquot, P. This, K. Chu, H. Johansson, and C. Meredith. 1999.

Historical genetics: the parentage of Chardonnay, Gamay, and other wine grapes of northeastern France. *Science* 285:1562–1565.

Bowers, J. E., and C. P. Meredith. 1997. The parentage of a classic wine grape, Cabernet Sauvignon. *Nature Genetics* 16:84–87.

THE PLANET'S MOST COMMON VERTEBRATE?

Miya, M., and M. Nishida. 1997. Speciation in the open ocean. *Nature* 389:803–804.

Nelson, J. S. 1994. *Fishes of the World*, 3rd ed. New York: Wiley.

12 Fossil DNA

ANCIENT ASPHALT JUNGLES

Harris, J. M., and G. T. Jefferson, eds. 1985. *Rancho La Brea: Treasures of the Tar Pits*. Los Angeles: Los Angeles County Natural History Museum.

Janczewski, D. N., N. Yukhi, D. A. Gilbert, G. T. Jefferson, and S. J. O'Brien. 1992. Molecular phylogenetic inference from saber-toothed cat fossils of Rancho La Brea. *Proceedings of the National Academy of Sciences USA* 89:9769–9773.

GENETIC VISIONS OF MAMMOTHS

Alroy, J. 2001. A multispecies overkill simulation of the end-Pleistocene megafaunal mass extinction. *Science* 292:1893–1896.

Greenwood, A. D., C. Capelli, G. Possnert, and S. Pääbo. 1999. Nuclear DNA sequences from late Pleistocene megafauna. *Molecular Biology and Evolution* 16:1466–1473.

Hagelberg, E., M. G. Thomas, C. E. Cook Jr., A. V. Sher, G. F. Baryshnikov, and A. M. Lister. 1994. DNA from ancient mammoth bones. *Nature* 370:333–334.

Höss, M., S. Pääbo, and N. K. Vereshchagin. 1994. Mammoth DNA sequences. *Nature* 370:333.

Noro, M., R. Masuda, I. A. Dubrovo, M. C. Yoshida, and M. Kato. 1998. Molecular phylogenetic inference of the woolly mammoth *Mammuthus primigenius*, based on complete sequences of mitochondrial cytochrome *b* and 12S ribosomal RNA genes. *Journal of Molecular Evolution* 46:314–326.

Ozawa, T., S. Hayashi, and V. M. Mikhelson. 1997. Phylogenetic position of mammoth and Steller's sea cow within Tethytheria demonstrated by mitochondrial DNA sequences. *Journal of Molecular Evolution* 44:406–413.

Pielou, E. C. 1991. *After the Ice Age*. Chicago: University of Chicago Press.

FACTS ON BRUIN EVOLUTION

Hänni, C., V. Laudet, D. Stehelin, and P. Taberlet. 1994. Tracking the origins of the cave bear (*Ursus spelaeus*) by mitochondrial DNA sequencing. *Proceedings of the National Academy of Sciences USA* 91:12336–12340.

Taberlet, P., and J. Bouvet. 1994. Mitochondrial DNA polymorphism, phylogeography, and conservation genetics of the brown bear *Ursus arctos* in Europe. *Proceedings of the Royal Society of London*, Series B, 255:195–200.

Talbot, S. L., and G. F. Shields. 1996. Phylogeography of brown bears (*Ursus arctos*) of Alaska and paraphyly within the Ursidae. *Molecular Phylogenetics and Evolution* 5:477–494.

THE DIETS OF SLOTHS

Hofreiter, M., H. N. Poinar, W. G. Spaulding, K. Bauer, P. S. Martin, G. Possnert, and S. Pääbo. 2000. A molecular analysis of ground sloth diet through the last glaciation. *Molecular Ecology* 9:1975–1984.

Höss, M., A. Dilling, A. Currant, and S. Pääbo. 1996. Molecular phylogeny of the extinct ground sloth *Mylodon darwinii*. *Proceedings of the National Academy of Sciences USA* 93:181–185.

Poinar, H. N., M. Hofreiter, W. G. Spaulding, P. S. Martin, B. A. Stankiewicz, H. Bland, R. P. Evershed, G. Possnert, and S. Pääbo. 1998. Molecular coproscopy: dung and diet of the extinct ground sloth *Nothrotheriops shastensis*. *Science* 281:402–406.

THE DEMISE OF THE FLIGHTLESS MOA

Clout, M. 2001. Where protection is not enough: active conservation in New Zealand. *Trends in Ecology and Evolution* 16:415–416.

Cooper, A., C. Mourer-Chauviré, G. K. Chambers, A. von Haeseler, A. C. Wilson, and S. Pääbo. 1992. Independent origins of New Zealand moas and kiwis. *Proceedings of the National Academy of Sciences USA* 89:8741–8744.

Craig, J., S. Anderson, M. Clout, B. Creese, N. Mitchell, J. Ogden, M. Roberts, and G. Ussher. 2000. Conservation issues in New Zealand. *Annual Review of Ecology and Systematics* 31:61–78.

Fleming, C. A. 1979. *The Geological History of New Zealand and Its Life.* Auckland, New Zealand: Auckland University Press.

Haddrath, O., and A. J. Baker. 2001. Complete mitochondrial DNA genome sequences of extinct birds: ratite phylogenetics and the vicariance biogeography hypothesis. *Proceedings of the Royal Society of London,* Series B, 268:939–945.

Herrmann, B., and S. Hummel, eds. 1993. *Ancient DNA.* New York: Springer-Verlag.

THE QUAGGA QUANDARY

Higuchi, R. G., B. Bowman, M. Freiberger, O. A. Ryder, and A. C. Wilson. 1984. DNA sequences from the quagga, an extinct member of the horse family. *Nature* 312:282–284.

Higuchi, R. G., L. A. Wrischnik, E. Oakes, M. George, B. Tong, and A. C. Wilson. 1987. Mitochondrial DNA of the extinct quagga: relatedness and extent of postmortem change. *Journal of Molecular Evolution* 25:283–287.

See also the Web sites: http://www.museums.org.za/sam/quagga/ quagga.htm; *and,* http://www.bbc.co.uk/qed/quagga.shtml.

NEANDERTHALS AND US

Krings, M., A. Stone, R. W. Schmitz, H. Krainitzki, M. Stoneking, and S. Pääbo. 1997. Neanderthal DNA sequences and the origin of modern humans. *Cell* 90:19–30.

Lewin, R. 1993. *Human Evolution: An Illustrated Introduction.* 3rd ed. Oxford: Blackwell.

Index